Dorit Jerger

Radar Forward Operator for Verification of Cloud Resolving Simulations within the COSMO Model

Wissenschaftliche Berichte des Instituts für Meteorologie und Klimaforschung des Karlsruher Instituts für Technologie (KIT)
Band 62

Herausgeber: Prof. Dr. Ch. Kottmeier

Institut für Meteorologie und Klimaforschung
am Karlsruher Institut für Technologie (KIT)
Kaiserstr. 12, 76128 Karlsruhe

Eine Übersicht über alle bisher in dieser Schriftenreihe erschienenen Bände finden Sie am Ende des Buches.

Radar Forward Operator for Verification of Cloud Resolving Simulations within the COSMO Model

by
Dorit Jerger

KIT Scientific Publishing

Dissertation, Karlsruher Institut für Technologie (KIT)
Fakultät für Physik, 2013
Referent: Prof. Dr. Klaus Dieter Beheng
Korreferent: PD Dr. Michael Kunz

Impressum

KIT Scientific Publishing

Karlsruher Institut für Technologie (KIT)
KIT Scientific Publishing
Straße am Forum 2
D-76131 Karlsruhe

KIT Scientific Publishing is a registered trademark of Karlsruhe Institute of Technology. Reprint using the book cover is not allowed.

www.ksp.kit.edu

Print on Demand 2014

ISSN 0179-5619
ISBN 978-3-7315-0172-5
DOI: 10.5445/KSP/1000038411

Radar Forward Operator for Verification of Cloud Resolving Simulations within the COSMO Model

Radarvorwärtsoperator zur Verifizierung von wolkenauflösenden Simulationen innerhalb des COSMO-Modells

Zur Erlangung des akademischen Grades eines

DOKTORS DER NATURWISSENSCHAFTEN

von der Fakultät für Physik des

Karlsruher Instituts für Technologie (KIT)

genehmigte

DISSERTATION

von

Dipl.-Phys. Dorit Jerger geb. Epperlein

aus Riesa

Tag der mündlichen Prüfung: 20. Dezember 2013

Referent: Prof. Dr. Klaus Dieter Beheng

Korreferent: PD Dr. Michael Kunz

Abstract

The German weather radar network comprises 17 C-Band dual polarisation Doppler radar systems evenly distributed throughout Germany for complete coverage. They provide unique information about precipitation in three dimensions and high temporal and spatial resolution. Up to now, these 3D data are not used within the numerical weather prediction COSMO model of DWD. This thesis contributed to the development of a radar forward operator for the COSMO model to derive synthetic radar measurements from the predicted model variables, in order to facilitate model verification (and data assimilation) using these data. These can be ways to improve the quantitative precipitation forecast based on numerical models.

This radar forward operator simulates the radar observables reflectivity and radial velocity. The model values are interpolated from the model grid onto the spherical coordinates of the radar system (azimuth, elevation, range). Several physical processes and geometric effects affecting radar measurements are considered, so as to make simulations and observations as comparable as possible. The focus of this thesis is the calculation of the reflectivity with the possibility of using full Mie theory (with or without attenuation of the radar pulse) or simpler Rayleigh approximation. Furthermore lookup tables are applied to drastically enhance the computational efficiency of the Mie calculations which are very costly otherwise.

In the second part of this thesis the operator has been used for verification of the operational cloud microphysical scheme which is applied for resolutions sufficiently high to permit explicit calculation of convection within the COSMO model. This has been done by comparing measured data with the simulated output of the radar forward operator by using and refining the method of CFADs.

Kurzfassung

Der deutsche Radarverbund besteht aus 17 C-Band Dopplerradargeräten ausgestattet mit Dual-Polarisationstechnik. Deutschlandweit liefern diese Geräte einzigartige, zeitlich und flächendeckend hochaufgelöste, dreidimensionale Informationen über Niederschlag. Bisher werden diese 3D-Daten jedoch kaum im COSMO-Model, dem numerischen Wettervorhersagemodell des Deutschen Wetterdienstes (DWD), verwendet. Diese Arbeit lieferte einen Beitrag zur Entwicklung eines Radarvorwärtsoperators für das COSMO-Model für die Herleitung von künstlichen Radarmessgrößen aus den prognostizierten Modellvariablen. Damit soll es für die Modellverifikation (und Datenassimilation) erleichtert werden, diese Daten zu verwenden. Die auf Modellrechnungen basierende quantitative Niederschlagsvorhersage kann dadurch verbessert werden.

Der Radarvorwärtsoperator simuliert die Radarmessgrößen Reflektivität und Radialgeschwindigkeit. Dabei werden die Werte vom Modellgitter in polare Radarkoordinaten (Azimut, Elevation, Radialentfernung) interpoliert. Bei der Simulation werden verschiedene physikalische Prozesse und geometrische Effekte berücksichtigt, um die somit erhaltenen Größen mit den Messwerten so vergleichbar wie möglich zu machen. In dieser Arbeit liegt der Fokus auf der Berechnung der Reflektivität. Dabei gibt es die Möglichkeit die vollständige Mie-Theorie (mit oder ohne Dämpfung des Radarstrahls) oder eine vereinfachte Rayleigh-Näherung zu verwenden. Zudem sorgen Lookup-Tabellen für eine deutliche Verbesserung der Rechenzeit für die ansonsten sehr kostenintensive Berechnung der Mie-Theorie.

Im zweiten Teil dieser Arbeit wurde der Operator in Hinblick auf die Verifikation von wolkenmikrophysikalischen Prozessen verwendet, bei einer Auflösung des COSMO-Modells, die es erlaubt Konvektion explizit zu berechnen. Dafür wurden Vergleiche von gemessenen Werten mit der simulierten Ausgabe des Radarvorwärtsoperators herangezogen. Die für den Vergleich verwendete CFAD-Methode wurde dabei genauer untersucht.

Contents

1 Introduction

Weather radar systems detect clouds and particularly precipitation at a high temporal and spatial resolution and is thus superior to any other measurement technique of these parameters. With its increasing deployment in the last decades resulting in a nationwide coverage of radar observables, they provide data that are useful, as e.g., for validation of ground-based precipitation measurements. Moreover, the analysis of radar data tracks of severe thunderstorms can be extracted (Handwerker, 2002) and damaging hail tracks can be investigated (Puskeiler, 2013). Radar data give unique information on temporal and spatial processes leading to clouds and precipitation. Consequently, since numerical weather prediction (NWP) models simulate the formation and development of clouds and precipitation, it is obvious that it should be able to derive radar products based on the output of appropriate fields of these water constituents (hydrometeors). Thus, hints and conclusions about the representation of atmospheric water particles within the model system can be drawn, which is an important step towards further improving microphysical schemes and their prediction.

The basic problem in combining measured with modelled radar data is that the quantities given by radar systems are not directly comparable to the model output. Firstly, the radar observables (reflectivity, radial velocity and polarisation parameters) do not represent specific variables of the model, like pressure, temperature, wind or the total mass fraction of the water constituents in the atmosphere. Secondly, the observed radar values are available on a coordinate grid related to the radar instrument that differs from the numerical grid of the model. The main objective of this project to be presented here is the provision of radar data for model verification. Note that in an accompanying project the main objective is the delivery of radar data for data assimilation (Zeng, 2013).

For both applications a so-called radar forward operator has been developed, a tool that calculates the radar observables (in the following reflectivity and radial velocity) from the predicted model variables and represents them in the same coordinate system as a 'real' radar would measure. The program module is embedded in the DWD's (Deutscher Wetterdienst, the national meteorological service of Germany) numerical weather prediction model that is called COSMO[1] model (formerly named "Lokal Modell" LM), and customised for the German radar network of DWD comprising 17 C band dual polarisation Doppler radar systems.

This radar forward operator differs from others (Lindskog et al., 2004; Caumont et al., 2006; Pfeifer et al., 2008) in its physical completeness and accuracy, especially with respect to the propagation of electromagnetic radiation in the atmosphere (describing the geometry of the radar beam) as well as the scattering by atmospheric targets (hydrometeors). Since the operator is going to be also used for operational data assimilation, additional effort is made to minimise the computational cost which is realised by simplifications/approximations as well as by explicit vectorisation and parallelisation of the program code. The complete forward operator has a modular design that also enables adjusting the balance between accuracy and efficiency as required.

This thesis is organised as follows: the COSMO model is briefly presented in chapter 2, the microphysics of clouds and precipitation in chapter 3 and the measuring technique of weather radar in chapter 4. The main topic of this work is divided into two parts. First, in chapter 5 the development of the radar forward operator is elaborated, with special emphasis on the simulation of the radar reflectivity. This includes the possibility of using the full Mie theory with or without extinction as well as the simpler Rayleigh approximation. Furthermore, lookup tables are calculated to speed-up the complex Mie calculation on

[1]Consortium for Small Scale Modelling

request. All these methods are discussed in great detail in the following. The second part (chapter 6) presents a number of examples with respect to model verification. In this regard, the results in form of contoured frequency by altitude diagrams (CFADs) are discussed, that compare vertically stratified areal-mean statistics.

It is finally stressed that a second PhD project, performed by Yuefei Zeng (2013), was intended in close cooperation to this work. At first, the development of the radar forward operator was distributed in equal parts between both projects. Afterwards, Y. Zeng focused on the application of the operator in data assimilation as mentioned above.

2 The COSMO Model

2.1 General Overview

The national meteorological service of Germany, Deutscher Wetterdienst (DWD), uses the COSMO model (Schättler et al., 2012; Baldauf et al., 2011), a non-hydrostatic, limited area atmospheric model, for operational numerical weather prediction (NWP) as well as for other scientific purposes. The first version of the model that had been developed at the DWD became operational in 1999, initially under the name of "Lokal Modell" (LM). Since then the further development of the model has been maintained by COSMO – Consortium for Small-scale Modelling – an organisation consisting of several meteorological services and other national organisations. Currently they are from the seven European countries Germany, Switzerland, Italy, Greece, Poland, Romania and Russia.

At present two variants of the COSMO model are used operationally at DWD, firstly the so-called COSMO-EU, which roughly covers the region of Europe, and secondly the higher resolution COSMO-DE for Germany and its neighbouring countries. The main parameters of both models are displayed in Table 2.1 and an overview of their respective domain is shown in Figure 2.1. Even though both variants are based on the same basic model code, some equations and

Table 2.1: Comparison of both COSMO model variants that run operationally at DWD.

name of the model	operational since	horizontal grid length [km]	grid size horizontal	grid size vertical	time step [s]	forecast range [h]
COSMO-EU	1999	7	665×657	40	40	78
COSMO-DE	2007	2.8	421×461	50	25	21

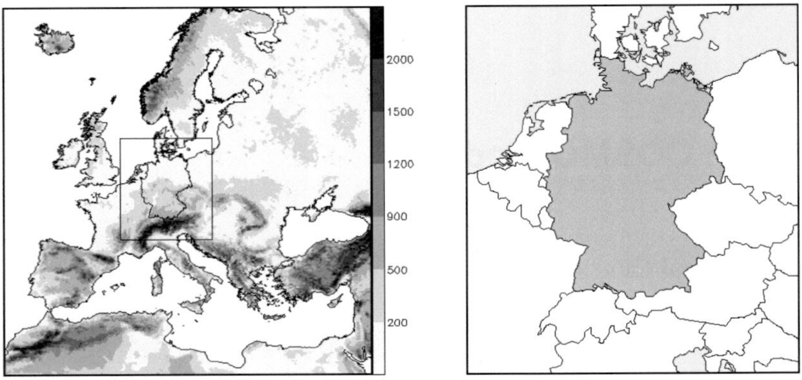

(a) (b)

Figure 2.1: (a) Domain of COSMO-EU in orographic view (colour bar in m a.s.l.). The framed box marks the embedded higher resolved COSMO-DE with the contours of Germany. (b) Domain of COSMO-DE in political view, with Germany highlighted.

model parameters given in this thesis may apply to COSMO-DE. This is true especially for the microphysical treatment of the parametrisation of clouds and precipitation used in the model (cf. section 3.3). Since the regional COSMO model is a limited area model, lateral boundary conditions are needed. For COSMO-DE they are provided by COSMO-EU and in turn COSMO-EU obtains its boundary data from the global model GME, that had also been developed at DWD. The initial data for each forecast are given by separate continous data assimilation cycles[2] for both COSMO variants.

2.2 Model Equations

The COSMO model uses non-hydrostatic, compressible, hydro-thermodynamical equations to quantify the atmospheric flow of dry and moist air (Doms, 2011). The temporal evolution is described by so-called prognostic equations, derived from Euler equations and Reynolds averaging of turbulent motion. The

[2]3h-forecasts started every 3h from the end state of the last forecast.

time-dependent prognostic variables are wind $\vec{v}$, pressure p, temperature T and mass fractions of the different constituents of atmospheric water q_x (which later will be unravelled more specifically). Additionally there are time-independent diagnostic variables, most important the total density of air ρ.

The governing equation for the three-dimensional wind $\vec{v} = \{u, v, w\}$ is obtained from the budget equation of momentum

$$\rho \frac{d\vec{v}}{dt} = -\nabla p + \rho \vec{g} - 2\vec{\Omega} \times (\rho \vec{v}) - \nabla \cdot \mathbf{T}. \tag{2.1}$$

The main forces are pressure gradient force ∇p, gravity acceleration $\vec{g}$, Coriolis force due to rotation of the earth with angular velocity $\vec{\Omega}$ and friction described by the (Reynolds) stress tensor $\mathbf{T}$.

The pressure tendency equation and the heat equation are

$$\frac{dp}{dt} = -\left(\frac{c_p}{c_V}\right) p \nabla \cdot \vec{v} + \left(\frac{c_p}{c_V} - 1\right) Q_h + \left(\frac{c_p}{c_V}\right) Q_m, \tag{2.2}$$

$$\rho c_p \frac{dT}{dt} = \frac{dp}{dt} + Q_h, \tag{2.3}$$

c_p and c_V are the specific heats at constant pressure and constant volume, respectively. Q_h is the diabatic heating (if positive) or cooling (if negative) and Q_m describes the impact of changes of the different q_x-terms in moist air. Both contributions, Q_h and Q_m, are neglected at present in the pressure equation.

The mass fraction of the water constituents $q_x \equiv \rho_x/\rho$ can be subdivided into the specific content of water vapour q_v, which is identical with the specific humidity, the specific content of liquid water q_l, namely cloud droplets and rain, and frozen water q_f, that can be cloud ice, snow, graupel or hail. The budget equations are combined in

$$\rho \frac{dq_x}{dt} = -\nabla \cdot \vec{J}_x + I_x , \tag{2.4}$$

$$\text{with} \quad \sum_x I_x = I_v + I_l + I_f = 0$$

$$\text{and} \quad \sum_x q_x = q_d + q_v + q_l + q_f = 1$$

$$\Downarrow$$

$$\rho \frac{dq_v}{dt} = -\nabla \cdot \vec{F}_v - (I_l + I_f) , \tag{2.5}$$

$$\rho \frac{dq_{l,f}}{dt} = -\nabla \cdot (\vec{P}_{l,f} + \vec{F}_{l,f}) + I_{l,f} . \tag{2.6}$$

I_x represent the sources and sinks resulting from phase changes and collision processes between the liquid and frozen water constituents. The conservation of theses terms connects the budget equations of q_v, q_l and q_f, respectively. $\vec{J}_x$ stand for diffusion fluxes, namely the precipitation fluxes $\vec{P}$ and the diffusive fluxes $\vec{F}$ of water vapour (index v) or non-vaporous water (index l,f) respectively. The equations (2.1)-(2.4) represent the prognostic equations that are used in the COSMO model, where (2.4) subdivides into (2.5) and (2.6), respectively.

Based on the equation of state of an ideal gas, the density of the gaseous atmospheric air ρ_{dv} (dry air and water vapour) can be described,

$$p = \rho_{dv} R T \quad \Rightarrow \quad p = \rho_{dv} (R_d q_d + R_v q_v) T , \tag{2.7}$$

where the total specific gas constant R is divided into the dry part of the air and the content of water vapour in the atmosphere. Hence, the diagnostic equations for the total density ρ of moist air and its liquid and frozen water constituents is derived, neglecting the small contribution of the finite (non-zero) volume of frozen and liquid particles themselves

$$\rho = \frac{p}{R_d \left(1 + \left(\frac{R_v}{R_d} - 1\right) q_v - q_l - q_f\right) T} . \tag{2.8}$$

Equation (2.8) together with the prognostic equations represents the total set of budget equations of the COSMO model.

2.3 Horizontal and Vertical Grid Structure

The COSMO model uses a spherical coordinate system analogue to the geographic coordinate system with longitude λ and latitude φ to define the horizontal grid structure on a spherical earth. However, for a mathematical description two numerical difficulties arise from the convergence of the meridians (lines of constant longitude) towards the poles. First, both poles form a mathematical singularity where all meridians collapse at one single point. Second, the grid structure is more strongly distorted the closer the computational domain is located to the poles. This distortion is maximum at the poles and becomes a minimum near the Equator. Since the COSMO model is a limited area model these problems can be avoided by rotating the coordinate system relative to the original geographic system. The principle is shown by following simplified chart:

original geographic		Geographic		geographic point of the		rotated
grid (long.,lat.)	$\Leftarrow$	North Pole	$\curvearrowright$	rotated north pole	$\Rightarrow$	grid
(λ_g, φ_g)		$90°\mathrm{N}, 0°\mathrm{W}$		$P_N = (\lambda_g^N, \varphi_g^N)$		(λ, φ)

The transformation is done by defining a point P_N as new northern pole of the coordinate system (transformation equations see Doms, 2011). The new prime meridian is defined as the meridian that runs through both the new and the original North Poles. In case of the coordinate system used in both COSMO model variants, P_N is located in the Pacific Ocean at $40°\mathrm{N}, 170°\mathrm{W}$ and hence the Equator of the rotated grid runs through the middle of Europe. For the COSMO-DE a $0.025°$-resolution of λ and ϕ results in a grid spacing of approximately 2.8 km.

The vertical grid structure of the COSMO model is defined by terrain-following coordinates. This means the model layers adapt to the contour of the orography, see Figure 2.2. Thus, the model orography provides – a numerically easy to handle – lower boundary condition. At upper levels the terrain-following layers gradually turn into layers of constant altitude or pressure. The transition takes place linearly and is completed at the interfacial height, which can be chosen arbitrarily. From the interfacial height to the top of the model domain the horizontal layers remain plane. This dual principle of terrain-following and constant vertical layers is a so-called hybrid coordinate system. In Figure 2.2 should one recognise that the single layers are not equidistant. The spacing reduces towards the surface in order to enable a better resolution of the important physical processes that take place in the planetary boundary layer. The elevation of each layer is denoted by ζ. Together with the rotated spherical coordinates the three-dimensional location of a grid point is defined by $(\lambda_i, \varphi_j, \zeta_k)$.

A characteristic of the grid structure used in the COSMO model is the distribution of the model variables on the grid points. Every grid point marks the centre of a grid box on which the all variables except of the three-dimensional wind components are defined. These are displaced by half a grid length in their corresponding direction. Figure 2.3 demonstrates this arrangement. The method is well-known as Arakawa C/Lorenz-grid and allows a more accurate mathematical description and minimises numerical errors.

Because of the limited coarse discretisation of the model grid structure in time and space, most physical processes can not explicitly be solved by the model, depending on the grid size and the magnitudes the physical processes happening, respectively. Therefore parametrisation schemes have to be devised in order to capture even the smallest scale processes (e.g. turbulence and radiation). Most importantly that applies to the description of the development of clouds and precipitation, which play a central role in this thesis and hence a separate chapter (3) is dedicated to this subject.

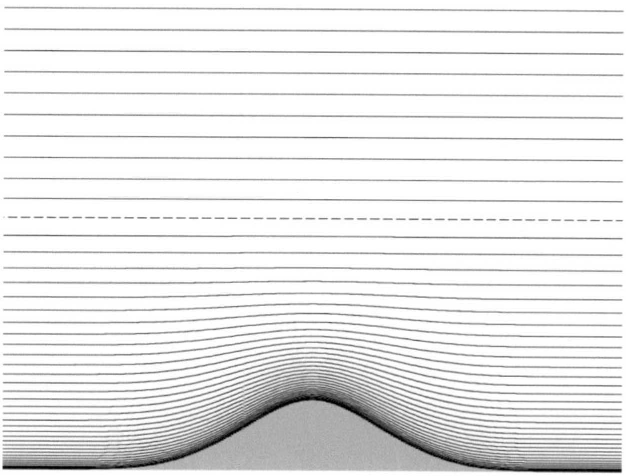

Figure 2.2: A two-dimensional display of the hybrid vertical model layers (in analogy to Steppeler et al., 2003) for an example of a simplified hill (filled grey). The interfacial height (dashed line) separates terrain-following layers from the layers of constant height coordinate (this can be altitude or pressure).

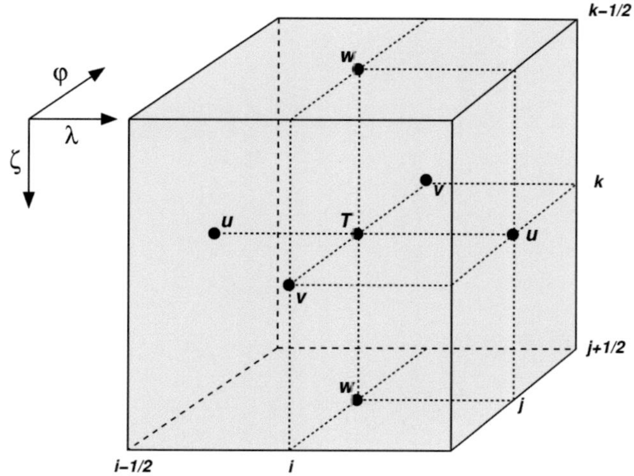

Figure 2.3: Arrangement of the model variables on an Arakawa C/Lorenz grid. The wind components are displaced by half a grid step. (Doms, 2011)

3 Microphysics of Clouds and Precipitation

3.1 Formation of Clouds and Precipitation

The atmosphere is a mixture of dry air and water, where water can exist in all states of matter: vapour, liquid and solid. The gaseous constituents are considered as ideal gases and the total density ρ can be described with the generalised equation of state (2.8) presented on page 8. The principal task of cloud microphysics is to describe all atmospheric processes leading to clouds and precipitation. In numerical weather prediction cloud microphysical schemes are used to parameterise the quantitative effects of these processes. The following explanations are mainly based on the descriptions from the textbook of Rogers (1979).

Water particles can be categorised according to their size, shape, formation, phase state as well as fall velocity, but some of these properties determine each other. However, their main distinctive features are their phase state and their characterisation as non-precipitating cloud constituents and those comprising precipitation. Other than precipitation particles, cloud particles are small enough to have an only small fall velocity compared to the surrounding ascending and descending air currents. These particles are either completely liquid (cloud water) or completely frozen (cloud ice). Precipitation particles are typically subdivided into rain and mixed-phase states (of ice/air or ice/(air/)water, when partially melted) like snow, graupel, and sometimes hail. A detailed description of all these different categories is given in section 3.3. Both cloud and precipitation particles are generally denoted as hydrometeors.

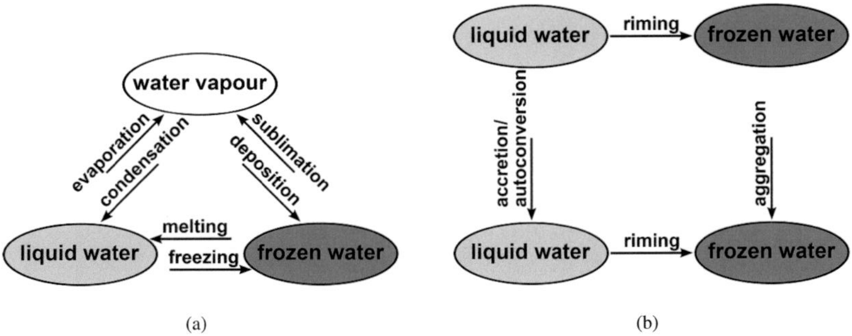

Figure 3.1: Microphysical processes which form clouds and lead to precipitation. (a) Phase changes between all three states of matter of water. (b) Collisional interactions between the liquid and the frozen state of water.

Clouds are formed by phase transitions. The initial process to form cloud droplets is nucleation from pure water vapour by condensation. This is called homogeneous nucleation, which requires several hundred percent saturation of relative humidity. For this reason it does not occur in the atmosphere. The more realistic process is heterogeneous nucleation that is the formation of water droplets on atmospheric aerosols, so-called cloud condensation nuclei. The same is also true for cloud ice, which can be produced either by freezing of supercooled cloud droplets, requiring ice nuclei, or by deposition of water vapour at deposition nuclei, respectively. Of course, all these phase transitions can also act in reverse, leading back to liquid water and water vapour, respectively, as shown in Figure 3.1(a).

In a warm cloud, where the temperature anywhere within the cloud is not colder than 0°C, only liquid water droplets exist. Once cloud water has formed the created cloud droplets grow either by diffusion of molecules of water vapour – in other words by further condensation – or by collection of slowly falling small water droplets by the faster larger ones. The efficiency of this collision-coalescence-process can be defined by the following expression:

$$\text{collection efficiency} = \text{coalescence efficiency} \times \text{collision efficiency}\,,$$

with the

$$\text{collision efficiency} = \frac{\text{number of collisions in path}}{\text{number of droplets in path}},$$

accounting for deflection in consequence of the surrounding current, and the

$$\text{coalescence efficiency} = \frac{\text{number of coalescence events}}{\text{number of collisions}}.$$

The collisional interactions between the primary cloud droplets are the dominant processes through which 'warm' precipitation develops. If growth continues the cloud droplets turn into raindrops that can reach the ground before they are completely evaporated below cloud base. If cloud droplets interact with each other to form raindrops the process is called autoconversion. The process of raindrops growing at the expense of cloud droplets is referred to accretion (Kessler, 1969).

If the temperature within the cloud drops below 0°C or the cloud top rises to altitudes above the freezing level besides supercooled droplets also cloud ice can occur. Cloud ice grows either by further deposition of water vapour or by collision interactions. In such a mixed-phase cloud interactions emerge between the liquid and the solid parts of water. Accumulating and instantaneous freezing of liquid water at the surface of frozen particles is called riming. The collision and sticking of frozen particles, mostly ice crystals, is called aggregation. Figure 3.1(b) demonstrates all possible interactions within a mixed-phase cloud.

3.2 Particle Size Distributions

For further considerations it is important to specify the size distribution functions $N(D)$ in terms of a size parameter, in this case the diameter D of the particles, for every type of hydrometeors. $N(D)dD$ indicates the number of drops per unit volume with diameters in the interval $[D, D+dD]$. Early experimental observations suggested an inverse exponential form of the number density distribution:

$$N(D) = N_0 \exp(-\lambda D)\,,$$

with

N_O : intercept parameter, which means a hypothetically virtual value of $N(D=0)$

λ : slope factor.

For raindrops Marshall and Palmer (1948) first proposed for the two free parameters:

$$N_{0,r} = 8000\,\mathrm{mm}^{-1}\mathrm{m}^{-3} \qquad \text{and} \qquad \lambda_r = 4.1R^{-0.21}\,\mathrm{mm}^{-1}\,.$$

with the inserted rainfall rate R in $\mathrm{mm\,h^{-1}}$ as new and the only free parameter left.

Because of the irregular shape of snow (and other frozen particles) an equivalent diameter conveniently replaces the real diameter, for example the corresponding diameter of a water drop when the snowflake is completely melted. Here, Gunn and Marshall (1952) gave the following approximation:

$$N_{0,s} = 3800R^{-0.87}\,\mathrm{mm}^{-1}\mathrm{m}^{-3} \qquad \text{and} \qquad \lambda_s = 2.55R^{-0.48}\,\mathrm{mm}^{-1}\,.$$

A second possibility to avoid the problem of characterising the specific shape of snow is the use of the particle mass instead of its diameter (cf. Seifert, 2002). And there is also a third possibility that defines a sphere surrounding an arbitrary shaped ice crystal (e.g. needles). The resulting bulk density is given by the mass of the particle divided by the equivalent volume of the surrounding sphere.

A more generalised function describing the size spectra of the particles is the gamma distribution

$$N(D) = N_0 D^{\mu} \exp(-\lambda D)\,,$$

with the spectral width or shape parameter μ as an additional free parameter (in case of rain $\mu = \mu_r$). The gamma distribution is often used for cloud particles, because of its modal shape and its ability to $N(D) \to 0$ for $D \to 0$. It is also appropriate for rain drops that are evaporating and sedimenting below cloud base.

The most generic distribution function, as estimated at present, is the generalised gamma distribution with four free parameters:

$$N(D) = N_0 D^{\mu} \exp(-\lambda D^{\nu}).$$

It is obvious that choosing $\nu = 1$ will lead back to the gamma distribution and both $\nu = 1$, $\mu = 0$ to the exponential form.

By means of the drop size distribution many other quantities can easily be calculated via the moments of $N(D)$, e.g.:

zeroth moment – number density (total number of particles per volume):

$$N_{tot} = \int_0^{\infty} N(D) dD, \tag{3.1}$$

third moment – equivalent to liquid water content[3] (mass of liquid water per volume):

$$L = \frac{\pi \rho_w}{6} \int_0^{\infty} N(D) D^3 dD. \tag{3.2}$$

sixth moment – radar reflectivity factor for the Rayleigh approximation of the scattering coefficient (cf. equation (5.3) on page 50).

The more moments of the distribution function are known the more free parameters can be identified.

3.3 Implementation in the COSMO Model

In the COSMO model the budget equations (2.1)-(2.4) specified in section 2.2 are used for the calculation of the prognostic variables. Describing the evolution

[3]The ice water content would be defined with the respective parameters for frozen water.

of the liquid and frozen water content in the atmosphere a so-called one-moment bulk water continuity model is applied (Doms et al., 2011). This means that the only parameter that is predicted is the total mass fraction q_x for various categories (index x) of hydrometeors. Therefore the size distribution for each type should have only one free parameter. Taking the exponential distribution into account, the slope factor λ_x can be related to q_x whereas N_0,x has to assumed to be a constant value or vice versa. Knowing q_x, the slope factor can then be determined by inverting the following integral

$$q_x = \frac{N_0,x}{\rho} \int_0^{\infty} m(D) \exp(-\lambda_x D) dD,$$

where ρ is the total density and $m(D)$ is the mass of the particle with diameter D. And with the assumptions on the size distribution other moments can be calculated.

In the COSMO-DE model the so-called three-category ice scheme or graupel scheme is in operational use. In this configuration the following categories are defined according to their phase state, shape, size, size distribution and fall velocity (for references see (Doms et al., 2011)):

non-precipitating water categories:

- water vapour (q_v) – gaseous aggregate state of water. The specific content of water vapour in the atmosphere is given by the budget equation (2.5) and provides the initial basis for the formation of all other water categories.
- cloud water (q_c) – the smallest liquid water droplets in the atmosphere with diameters up to $\sim 100\,\mu$m. They have a negligible terminal fall velocity and are considered as spheres. In the COSMO model cloud water has no specific size distribution.
- cloud ice (q_i) – the smallest frozen water particles in the atmosphere with a pristine crystal structure (small hexagonal plates with $D_i \lesssim 200\,\mu$m) and a negligible terminal fall velocity as well. The

size distribution is monodisperse (defined by a δ-peak) and the temperature-dependent number density, related to the freezing point $T_0 = 273.15\,\mathrm{K}$, is given by

$$N_i(T) = N_{0,i}\exp(0.2(T_0 - T)) \quad \text{with} \quad N_{0,i} = 1.0 \times 10^2\,\mathrm{m}^{-3}\,.$$

- crystal mass: $m_i = \rho q_i N_i^{-1}$
- mass-size-relation: $m_i = a_{m,i} D_i^3$ with form factor $a_{m,i} = 130\,\mathrm{kg\,m}^{-3}$

precipitating water categories:

- rain (q_r) – liquid water drops with a diameter of $\gtrsim 100\,\mu\mathrm{m}$ up to 5 mm. They are considered as spheres (disregarding that falling larger drops are rather oblate).
  - gamma distribution with $N_{0,r} = 6 \times 10^4 \exp(3.2\mu_r)$ (Ulbrich, 1983)
  - water density: $\rho_w = 1000\,\mathrm{kg\,m}^{-3}$
  - mass-size-relation: $m_r = a_{m,r} D_r^3$ with form factor $a_{m,r} = \pi\rho_w/6\,\mathrm{kg\,m}^{-3}$
  - terminal fall velocity: $v_{T,r}(D_r) = v_{0,r} D_r^{0.5}$ with $v_{0,r} = 130\,\mathrm{m}^{0.5}\,\mathrm{s}^{-1}$
- snow (q_s) – completely frozen aggregates of ice crystals of various shapes and comparatively small bulk density.
  - exponential size distribution with the temperature-dependent parameter $N_{0,s}(T)$ according to Field et al. (2007)
  - mass-size-relation: $m_s = a_{m,s} D_s^2$ with form factor $a_{m,s} = 0.38\,\mathrm{kg\,m}^{-2}$
  - terminal fall velocity: $v_{T,s}(D_s) = v_{0,s} D_s^{0.25}$ with $v_{0,s} = 4.9\,\mathrm{m}^{0.75}\,\mathrm{s}^{-1}$

- graupel (q_g) – ice category with higher bulk density and higher fall velocity than snow, considered spheres.
  - exponential size distribution with $N_{0,g} = 4 \times 10^6\,\mathrm{m}^{-4}$ (Rutledge and Hobbs, 1984)
  - mass-size-relation: $m_g = a_{m,g} D_g^{3.1}$ with form factor $a_{m,g} = 169.6\,\mathrm{kg\,m}^{-3.1}$
  - terminal fall velocity: $v_{T,g}(D_g) = v_{0,g} D_g^{0.89}$ with $v_{0,g} = 442.0\,\mathrm{m}^{0.11}\,\mathrm{s}^{-1}$
- hail (q_h) – ice category with very high bulk density and large terminal fall velocity. Only used in an alternative non-operational two-moment scheme (Seifert and Beheng, 2006; Noppel et al., 2006).

A more detailed scheme is the two-moment parametrisation, which has a second prognostic variable per hydrometeor category. In the COSMO-DE model this is the number density N_x (3.1) besides the mass fraction q_x (Seifert and Beheng, 2006). Also an additional water category, namely hail, is included (Noppel et al., 2006). This scheme allows for another free parameter for the size spectrum but requires more computing time and is only used for case studies and research applications to date.

To conclude this chapter, Figure 3.2 summarises all water categories and microphysical processes which are parameterised in the one-moment three-category ice scheme in the COSMO-DE model. The presented scheme has been applied in the framework of this thesis.

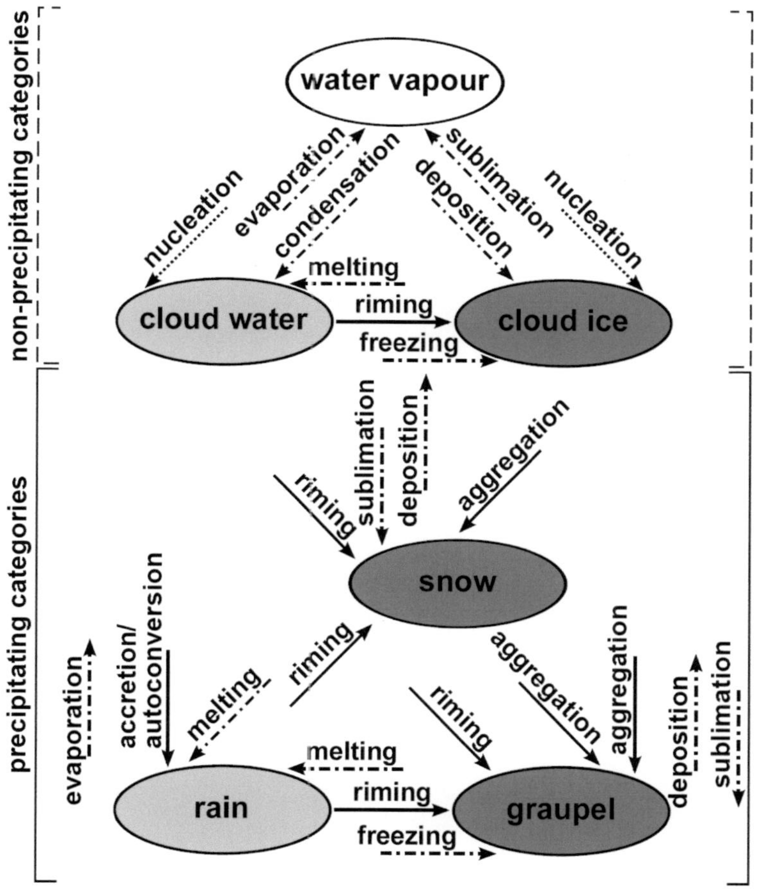

Figure 3.2: The one-moment bulk water-continuity model and three-category ice scheme or graupel scheme that is implemented in the COSMO model with six overall water categories (including three different types of frozen water substance). The grey scales indicate the aggregate state analogue to Figure 3.1. The microphysical processes are either phase transitions as in 3.1(a) indicated by the dash-dotted lines – or interactions as in 3.1(b) indicated by the solid lines.

4 Dual Polarisation Doppler Weather Radar

4.1 Principle of Measurement

The previous chapter focused on the theoretical description of clouds and precipitation, in this chapter an important measurement technique for these atmospheric characteristics is described which is provided by weather radar systems. These systems detect clouds and precipitation in short time intervals from remote on a long range of over hundred kilometres and thus provide an essential tool for precipitation analysis.

A radar uses **ra**dio waves (microwaves) in order to provide the **d**etection **a**nd **r**anging of present atmospheric objects, hydrometeors at its best, and moreover to determine their characteristics. For this purpose, a short pulse of electromagnetic radiation with well-known frequency, intensity, duration (length) and direction is sent into the atmosphere by a parabolic (in case of weather radar) antenna. If the pulse hits a target whose refractive index differs from that of air, the incoming wave is scattered and is partly reflected to the radar. The backscattered signal is called an "echo", and by the amount conclusions are drawn from it about the existence, quantity and size of the scattering objects. Depending on the configuration of the radar system, conclusions about the motion, shape and state of matter of the different hydrometeor types can be made in addition. Radars are widely used in meteorology and can be applied for different purposes. Their range of applicability for certain phenomena is determined by the frequency (or wavelength) which is used, due to the strong variation of its influence on attenuation (see paragraph 5.4.3). Radio frequencies are divided into so-called "bands". Table 4.1 shows the frequency bands of the radio spectrum

Table 4.1: List of the radar frequency bands according to the IEEE standard.

Band	Frequency [GHz]	Wavelength [cm]	Usage in meteorology
L	1-2	15-30	temperature measurement via radio acoustic sounding systems
S	2-4	7.5-15	long-range detection of (heavy) precipitation
C	4-8	3.75-7.5	medium-range detection of precipitation
X	8-12	2.5-3.75	short-range detection of (clouds and) precipitation
K_u	12-18	1.67-2.5	
K	18-27	1.11-1.67	detection of clouds
K_a	27-40	0.75-1.11	detection of clouds
V	40-75	0.4-0.75	
W	75-110	0.27-0.4	high-resolution, short-range detection of clouds

as specified by the Institute of Electrical and Electronics Engineers (IEEE) and their most common usage in meteorology. Most weather radars in Europe operate at C band. The following considerations are mainly based on the textbooks of Sauvageot (1992) as well as Doviak and Zrnic (1993).

Electromagnetic radiation propagates at the speed of light (in a horizontally stratified medium) and hence from the elapsed time before an echo is received the range r of the scatterers can be identified. The maximum detectable range is determined by the pulse repetition frequency (PRF), the number of pulses per second. The antenna transmits the pulse, forming the so-called radar beam, and from the orientation of the antenna (in addition to the arrival time of an echo) the precise location of the scatterer in terms of (r, ϕ, θ) is known. Here, the orientation of the antenna is given by the azimuth angle ϕ, that is the angle between the radar beam and a vector pointing north projected onto the surface of the earth, and the elevation angle θ, the angle between the radar beam and the surface of the earth.

However, the radar pulse broadens with increasing distance and hence the received signals are from larger and larger volums. In order to describe the broadening a second coordinate system is introduced denoted by the angles θ_p and ϕ_p, respectively, with respect to the centre of the radar pulse. A sketch of a single radar pulse propagating along the radar beam is shown in Figure 4.1 with the corresponding coordinate systems. Because of the beam broadening, scatterers from an ever enlarging volume contribute to the received signal from still one single radar pulse. This is a main problem which will be discussed in more detail later on.

Weather radars are typically used as volume-sensing devices. During the measurement of consecutive pulses, the antenna rotates azimuthally through 360 degrees once, then switches the elevation and rotates again. This procedure is repeated until the final elevation is reached such that then an entire volume scan is finished, then it starts again with the first (in most cases lowest) elevation. Thereby measurements of the atmospheric scatterers are produced inside the volume of a cone. Figure 4.2 shows the resulting coverage of such a scan strategy. In that way a three-dimensional and comprehensive atmospheric measurement is provided and not only the precipitation reaching the ground but moreover its formation and growth in the atmosphere can be detected. This is one of the significant advantages of the radar technique.

Directivity and Radiation pattern

As stated before, in order to obtain a localisation as precise as possible the antenna sends a directed and preferably narrow pulse. However, due to technical constraints there are two restrictions that are inevitable and hence have to be considered. As was already mentioned, the transmitted radar pulse widens with distance both in azimuthal (ϕ_p) and in elevational (θ_p) direction. As the antenna dish is usually circular as for meteorological radars, this effect is symmetrical in both directions. Furthermore, the transmitted power density has in a good approximation a Gaussian shaped angular distribution and therefore the power distribution does not look like a cone as illustrated in Figure 4.1 but more like a "lobe" as can be seen in Figure 4.3. The picture also shows, that the antenna

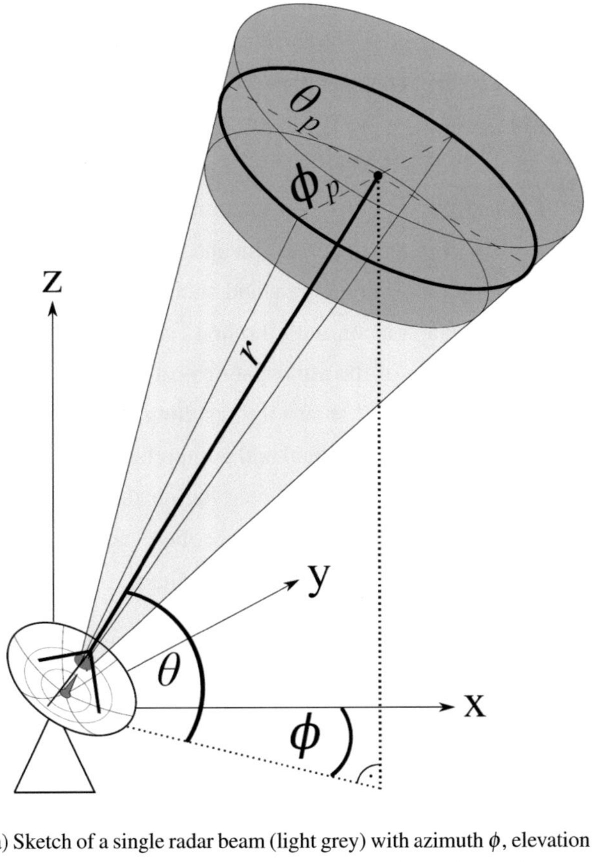

Figure 4.1: (a) Sketch of a single radar beam (light grey) with azimuth ϕ, elevation θ and range r (related to a Cartesian coordinate system x, y, z) pointing to the centre of the radar pulse (dark grey section), that broadens with distance described by the coordinates ϕ_p and θ_p respectively.

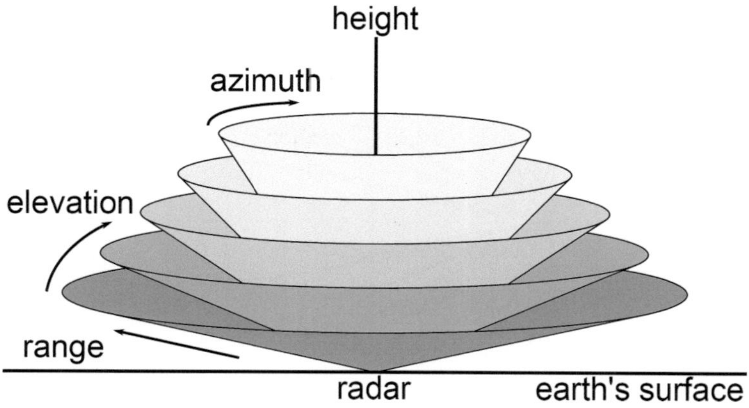

Figure 4.2: Simplified illustration of "all" radar beams composing a so-called volume scan (in analogy to Ruffieux and Illingworth, 2013). Here each grey scale indicates all azimuth scans of one single elevation step.

unfortunately does not only emit its power in the forward direction only but in smaller parts also sideways and even backwards in additional lobes, called side lobes and back lobe respectively. The main lobe contains the major part of the power. In order to give a quantitative description of this effect there are two parameters that specify the directional characteristic.

First, the direcivity $D(\theta_p, \phi_p)$ is a measure of the capability of the antenna to send the transmitted power P_t in one specific direction with power density S_{dir} relative to a source that radiates isotropically with power density S_{iso}. So, at any distance r it is (here, first of all without attenuation)

$$S_{iso}(r) = \frac{P_t}{4\pi r^2} \quad \text{and} \quad S_{dir}(r, \theta_p, \phi_p) = \frac{P_t}{4\pi r^2} D(\theta_p, \phi_p) \tag{4.1}$$

$$\Rightarrow \quad D(\theta_p, \phi_p) = \frac{S_{r,dir}(\theta_p, \phi_p)}{S_{r,iso}}.$$

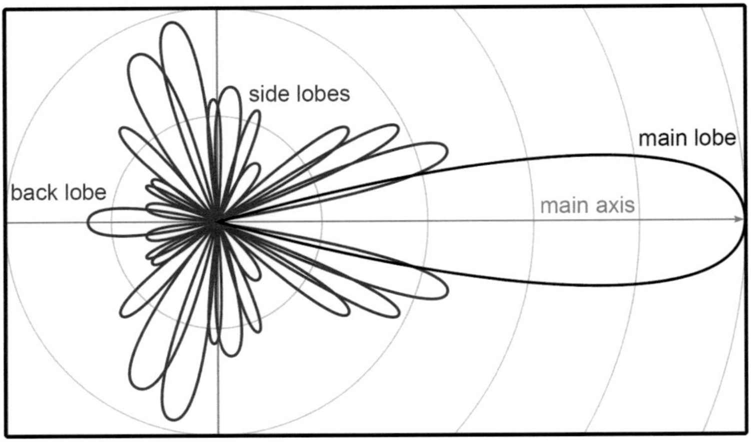

Figure 4.3: Power distribution of a radar beam known as radiation pattern on a polar grid in logarithmic scale (in analogy to Sauvageot, 1992).

For further considerations it is more convenient to use the normalised intensity or radiation pattern $f^2(\theta_p, \phi_p)$. Given the maximum intensity on the beam axis, this means at $D_0 = D(\theta_p = 0, \phi_p = 0)$, the radiation pattern is defined by

$$f^2(\theta_p, \phi_p) = \frac{D(\theta_p, \phi_p)}{D_0} = \frac{S_{r,dir}(\theta_p, \phi_p)}{S_{r,dir}(0,0)}$$

It should be mentioned that these definitions are only valid in the far-field region of the source of radiation, which roughly starts at a radial distance of $2d^2/\lambda$ where d is the diameter of the antenna dish and λ the wavelength of the emitted radiation. The scattering signals from objects that are located closer to the radar are disregarded by the radar software.

Radar equation

A weather radar transmits the emitted power P_t and measures the received power P_r which depends on a few radar parameters but most importantly on the properties of the scattering objects, in other words the backscattering cross-section σ_b (its detailed definition is given in paragraph 5.4.2). Via the radar equation a relation between these quantities is created. Starting with the assumption of there being only a single particle that scatters and as an extension of equation (4.1) for a directional antenna, the relation has the following gross form

$$P_r = P_t C_{radar} \left(\frac{f^2}{r^2}\right)^2 \frac{1}{l^2} \sigma_b .$$

C_{radar} is a technical constant of the specific radar system, including the wavelength and some properties of the antenna, and l^{-2} is the part of the damping of the radar pulse during its propagation to the position of the target. The squared terms originate from both the path lengths from the radar to the scatterer as well as the distance from the scatterer back to the radar.

The next step is to extend the foregoing consideration to a large number of scatterers. Thus the radar measures at a specific time the sum of the signals of all particles that are located in the pulse resolution volume[4]. Since nobody knows exactly how the particles[5] are distributed in such a volume, a uniform distribution is presumed with diameters according to an assumed size spectral function $N(D)$ which was introduced in section 3.2. Consequently the radar equation for the averaged received power from a size-distributed ensemble of scatterers is composed as follows

$$\overline{P_r} = P_t C_{radar} \int\limits_{r_0-\delta/2}^{r_0+\delta/2} \int\limits_{-\pi}^{+\pi} \int\limits_{-\pi/2}^{+\pi/2} dV \frac{f^4(\theta_p, \phi_p)}{r^4} |W(r)|^2 \frac{1}{l^2(r, \theta_p, \phi_p)} \int\limits_0^\infty dD \, \sigma_b(D) N(D, r, \theta_p, \phi_p) , \tag{4.2}$$

with $dV = dV(r, \phi_p, \theta_p) = r^2 \cos\theta_p d\theta_p d\phi_p dr$. In radial direction the integral is limited by the range δ of the radar pulse. In angular direction the contribution of the term is determined by f^2 which is typically approximated by a Gaussian function that has a strong slope towards its boundaries (which is truncated at some point). Additionally an range weighting $W(r)$ is applied, which can be assumed by a simple boxcar function.

So far it was tacitly supposed that only one particle type is present in the beam volume. In case of different hydrometeors (as in mixed-phase clouds) the

[4] We think of it as being the main lobe.

[5] supposing these are all hydrometeors

received signal is assumed to be the sum over all corresponding backscattering cross-sections and hydrometeor characteristics in the particle size distributions. It should be further noted that one single value represents the whole volume of the radar pulse at distance r, and that the assumption of a homogeneous coverage of the measuring volume with particles is just a rough approximation to reality. This causes a certain potential of misinterpretation of radar signals and has to be kept in mind.

4.2 Radar Observables

4.2.1 Radar Reflectivity

The inner integral over the particle diameter in equation (4.2) is the sum of all backscattering cross-sections of all particles per volume,

$$\eta = \int_0^\infty dD\, \sigma_b(D) N(D)\,. \tag{4.3}$$

This variable is referred to as radar reflectivity η. With the calculation of η (for σ_b, cf. paragraph 5.4.2) and the other aforementioned approximations, equation (4.2) can be simplified[6]

$$\overline{P_r} = \widetilde{C}_{radar} \frac{Z}{r^2 l^2}\,.$$

The quantity Z appearing in this relation is the so-called radar reflectivity *factor* (as the term radar reflectivity is already taken), an extension of η that is more convenient to work with. The reason for that will be described later. In addition, some more assumptions have to be made to define $\widetilde{C}_{radar}$ (e.g. all hydrometeors are raindrops) which leads to a further definition, the so-called *equivalent* radar reflectivity factor Z_e. The exact calculation of Z and Z_e, respectively, is given in paragraph 5.4.2 in great detail. The equivalent radar reflectivity factor is the quantity of interest for meteorologists working with radar data. It is a measure

[6] A derivation can be found e.g. in Sauvageot (1992).

of the number of particles per volume and a function of the size and the dielectric properties of the targets and hence gives information about the particles occurring in the atmosphere.

Note that in meteorological applications, the rate of rainfall R, commonly expressed in units of mm/h, is derived from Z_e. However, in this study the concentration is on the calculation of the radar reflectivity. It is assumed that a reliable relation between R and Z_e exists.

4.2.2 Radial Velocity

Besides the received intensity, most modern radar systems are able to measure the phase shift of the received signal from pulse to pulse such that the radial velocity of the scatterers can be obtained from the Doppler shift. The phase φ and hence the phase shift or Doppler shift $d\varphi/dt$ are given by the following relations:

$$\varphi(t) = 2\pi\frac{2r}{\lambda} \quad \text{and} \quad \frac{d\varphi}{dt} = 2\pi\frac{2v_r}{\lambda},$$

$$\text{with} \quad f_D = \frac{2v_r}{\lambda}.$$

The Doppler frequency f_D is the change in frequency due to the velocity of the hydrometeors towards or away from the radar with the radial velocity v_r. By convention a positive velocity ($v_r > 0$) results if the target is receding and a negative velocity ($v_r < 0$) if the target is approaching the radar. All other components of the velocity that are transverse to the direction of the radar beam cannot be measured. Nevertheless, these data can be interpreted[7] as additional information about the horizontal wind field. However, the presence of atmospheric objects is necessary, and without clouds and precipitation no radar measurement of the radial wind is possible. It should be further mentioned that the measured radial velocity of a radar pulse represents the (reflectivity- and beam function-)weighted average of the motion of all hydrometeors in the pulse volume.

[7] with the usual difficulties

A significant drawback in the measurement of the radial velocity is that only a phase shift within $\pm\pi$ is clearly detectable. At phase differences that exceed $\pm\pi$, the resulting v_r is folded around $-v_r$ and lead to an ambiguity. Thus, the received signal will be interpreted to a velocity which is shifted by multiples of $\pm 2v_r$. This effect is called aliasing and can be corrected afterwards with some effort and success. The maximum unambiguous velocity at an exact phase shift of π is called the Nyquist velocity.

4.2.3 Polarisation Parameters

Conventional Doppler radar systems measure the reflectivity of the hydrometeors and their motion relative to the radar. Since most atmospheric particles are not spheres – raindrops are oblate and snow can have any shape – it would be desirable to further know the exact shape (axis ratio) and additionally the state of matter of these particles. This can be achieved with the use of the dual polarisation technique, which means that the radar pulse can be transmitted and received at two polarisations, horizontally and vertically. Here, the corresponding reflectivity would be denoted by Z_{HH}, Z_{VH} and Z_{VV}, respectively, where the first index indicates the received and the second index indicated the transmitted polarisation direction. Thereby a set of new parameters can be defined:

Linear depolarisation ratio $LDR = Z_{VH}/Z_{HH}$

Differential reflectivity $Z_{DR} = Z_{HH}/Z_{VV}$

Differential phase shift $\Phi_{DP} = \Phi_H - \Phi_V$

Specific differential phase $K_{DP} = \Delta\Phi_{DP}/(2\Delta r)$

Correlation coefficient ρ_{hv}

These quantities depend on the angular orientation (canting angle) of the hydrometeors, their state of matter as well as its concentration (many small droplets or a few large drops). In combination they allow an interpretation about the shape (asphericity) and type of the scattering particles. Furthermore, the rate of rainfall can be derived more specifically. However, polarisation parameters are not further used in this thesis and here only briefly mentioned. A detailed describtion of polarimetric radars can be found in Bringi and Chandrasekar (2001).

4.3 Errors of Measurement

Besides the already mentioned assumptions made for interpretation and general uncertainties arising from a radar measurement (calibration, noise, etc.), there is a list of other typical errors coming along with the radar technique that have to considered. The following survey itemises possible sources of errors that are known so far. However, this list does not claim to be complete.

- The "bright band": Melting particles have a stronger backscattering signal than completely liquid or completely frozen hydrometeors, which could lead to an overestimation of the rate of rainfall in the presence of melting snow or graupel. On the other hand this effect helps to detect the melting layer of clouds, due to a zone of higher reflectivity values, that is known as the "bright band".

- Non-meteorological clutter (clear-air echoes): Clutter are unwanted radar echoes. Besides cloud and precipitation particles there are a lot of other objects that could scatter, like planes, birds, insects, dust or even pollen. Also atmospheric turbulence (caused by variations in the air refractive index) can lead to radar echoes. Usually these signals are very small and mostly seen on days without precipitation in the vicinity of the radar.

- Ground clutter: Clutter arising from the obstacles at ground is a special case of non-meteorological clutter. These can be reflexions from trees and other vegetation, buildings and other man-made constructions, orographic structures like rocks, hills and mountains or sea clutter. These mostly fixed obstacles as mountains may also lead to a (partial) shielding of the radar pulse.

- Second-trip echoes: Precipitation events that are further away from the radar than the maximum detectable range could not "answer" the first pulse before the second pulse is sent. So, the detected echo of the first pulse is mismatched to the second and hence the event is interpreted to be located closer than it is actually. Second-trip echoes can be sometimes recognised as azimuthal stripes in the graphics created from the radar data.
- Anomalous propagation of the radar beam (anaprop echoes): The propagation of electromagnetic radiation depends on the refractive index of the atmosphere. Under special atmospheric conditions it can deviate strongly from the "normal" case. The beam is, e.g., bent towards the earth's surface and ground clutter can occur at points where it normally does not (see section 5.2).
- Partial or non-homogeneous filling of the detecting volume of the radar pulse leads to a misinterpretation of the derived parameters like the precipitation intensity.
- Attenuation of the radar pulse due to atmospheric gases and hydrometeors: Absorption and side scatter of energy out of the pulse diminish the energy and reduces the echoes from far away (see paragraph 5.4.3).
- Scattering from the back or the side lobes, respectively.
- Aliasing: folding of the radial wind (see paragraph 4.2.2).

4.4 The Radar Network in Germany

The first radar station launched by the German Meteorological Service (Deutscher Wetterdienst, DWD) was installed in 1987 near Munich and was able to measure only the reflectivity. Gradually, 15 more radar stations were built and additionally equipped with Doppler capability until 2004. From 2010 on the already existing facilities are in the process of being replaced by state-of-the-art dual polarisation Doppler weather radars DWSR5001/C/SDP/CE from the Enterprise Electronics Corporation (EEC), which was done within the framework of the project RadSys-E (Radar-System-Ersatz). This project is planned to be finished in 2014. The new radar network is extended by an additional site in Memmingen and will finally comprise 17 radar stations altogether throughout Germany, each endowed with the capability to measure reflectivity, radial wind and polarisation parameters. Table 4.2 lists the names of all 17 radar stations, the year of the initial operation and the altitude of the antenna. In Figure 4.4 the corresponding locations within Germany can be seen. It shows that the network provides a complete coverage and can provide three-dimensional and high-resolution precipitation measurements for the whole country. A complete volume scan performed every 15 minutes comprises 18 elevations from 0.5° to 37.0°, in 1° azimuthal and 1 km range steps. The maximum range is 124 km and the elevations supply a vertical coverage up to approximately 12 km. Hence all measuring grid points of a single radar composing one volume scan are given by:

$$
\begin{aligned}
r \in & \{1.0, 2.0, 3.0, \ldots, 124.0\}\,\mathrm{km} \\
\phi \in & \{0, 1, 2, 3, \ldots, 359\}^{\circ} \\
\theta \in & \{0.5, 1.5, 2.5, 3.5, 4.5, 5.5, 6.5, 7.5, 8.5, 9.5, \\
& 11.0, 13.0, 15.0, 17.0, 19.0, 23.0, 29.0, 37.0\}^{\circ}.
\end{aligned}
$$

Table 4.2: Radar stations of the DWD.

location	since	altitude [m a.s.l.]
Boostedt	1990	124
Rostock	1995	36
Emden	1994	58
Hannover	1994	81
Ummendorf	1996	183
Prötzel	1991	189
Essen	2011	185
Flechtdorf	1997	623
Dresden	2000	262
Neuhaus	1994	879
Neuheilenbach	1998	585
Offenthal	2011	246
Eisberg	1997	799
Türkheim	1998	764
Schnaupping	1987	724
Feldberg	1997	1517
Memmingen	2011	720

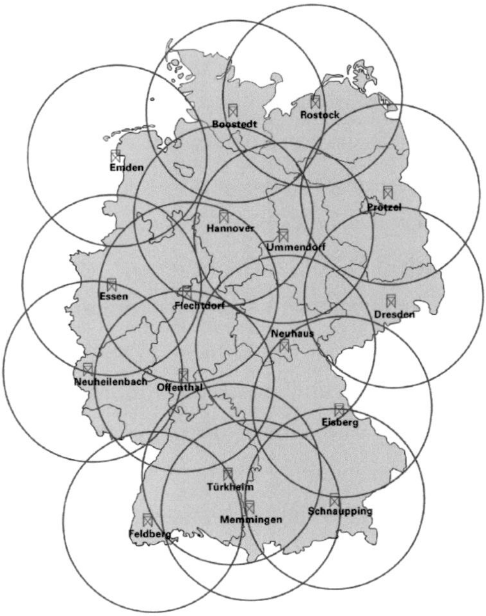

Figure 4.4: Map of the radar stations located in Germany.

5 Radar Forward Operator

5.1 General Structure

Data from radar measurements give essential information about the existence and quantity of hydrometeors in the atmosphere and can help to improve the numerical precipitation forecast. For this reason it is intended to make use of these data for data assimilation and model verification. However, as described in section 2.2 and 2.3 the output of the COSMO model are the prognostic variables given on a rotated geographical coordinate system as well as on terrain-following vertical coordinates, analogue to a nearly Cartesian coordinate system. On the contrary the weather radar measures its observables in the spherical coordinates azimuth, elevation and range related to the radar location (see section 4.1 and 4.2).

- The prognostic variables of the model are, with dependence on (x, y, z):
  - horizontal and vertical wind components: u, v, w
  - pressure: p
  - temperature: T
  - specific humidity: q_v
  - total mass fraction of cloud water and cloud ice: q_c, q_i
  - total mass fraction of rain, snow and graupel: q_r, q_s, q_g
- The radar observables are, with dependence on (r, ϕ, θ):
  - radar reflectivity: Z_e

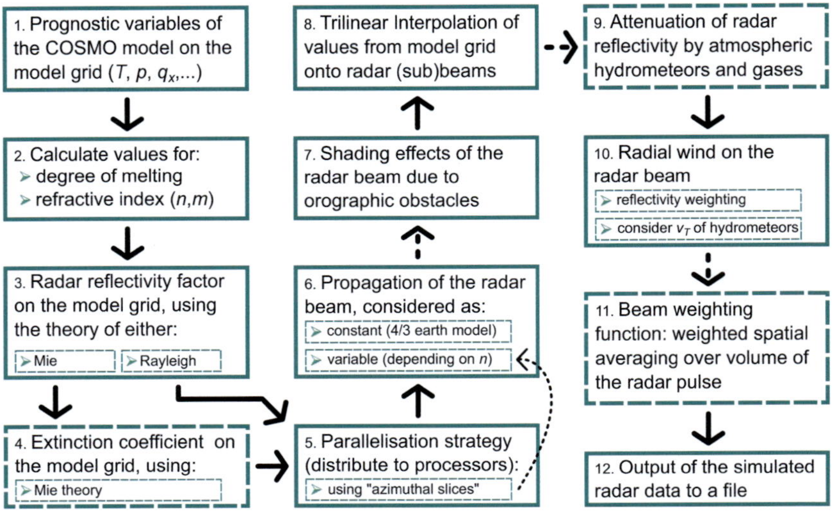

Figure 5.1: Conceptual flow chart of the full comprehensive and modular radar forward operator including all important physical processes. Dashed frames indicate optional modules.

- radial wind: v_r
- polarisation parameters: LDR, K_{DP}, Z_{DR}, ρ_{hv}

It is obvious that both types of variables are not directly comparable with each other. In order to permit quantitative comparisons they have to be converted into each other. For this purpose a radar forward operator has been developed in this investigation, a tool that simulates radar data, reflectivity and radial velocity[8], in the "forward direction" meaning from model output to observational data. The program is embedded in the COSMO model.

The operator was intended and designed at full complexity taking all physical aspects – that are known to have an impact on the measurement – into account in order to get the best possible accuracy. Concurrently, simplifications and approximations were implemented that – if required – can be switched on to

[8]Polarisation parameters are not considered so far.

enable a more efficient simulation. This was realised by a modular design. In Figure 5.1 a schematic overview of all individual operations which are necessary to create a complete simulation program of radar data is shown. Starting from the prognostic model data given on the model grid (1), the output are finally the simulated radar observables given in the coordinate system of the radar beam (12). Boxes (2)-(11) in between will be described in detail in the following sections: First, section 5.2 discusses steps (6), (7) and (11). Section 5.3 deals with step (10). The main focus of this work was the simulation of the reflectivity, thus section 5.4 elaborates boxes number (2), (3), (4) and (9). And some final remarks to part (5) of the flow chart will be given in section 5.7 to complete the radar forward operator. The development took place in close cooperation with Yuefei Zeng (2013), whose work will be referred to in a number of cases. His main contribution is described in the sections 5.2 and 5.7.

5.2 Propagation of the Radar Beam

Refraction

When electromagnetic waves propagate through a medium the speed of propagation c differs from the speed of light in vacuum c_0 by a factor n, the refractive index, such that $c_0 = c \cdot n$, depending on the type of medium. In vacuum this factor is unity, in any other environment it is always $n > 1$, i.e. the speed of radiation is limited by the medium (the greater n, the smaller c) which is penetrating. The refractive index of atmospheric air differs only slightly from vacuum. So, for convenience the so-called refractivity N is introduced:

$$N = (n-1)10^6$$

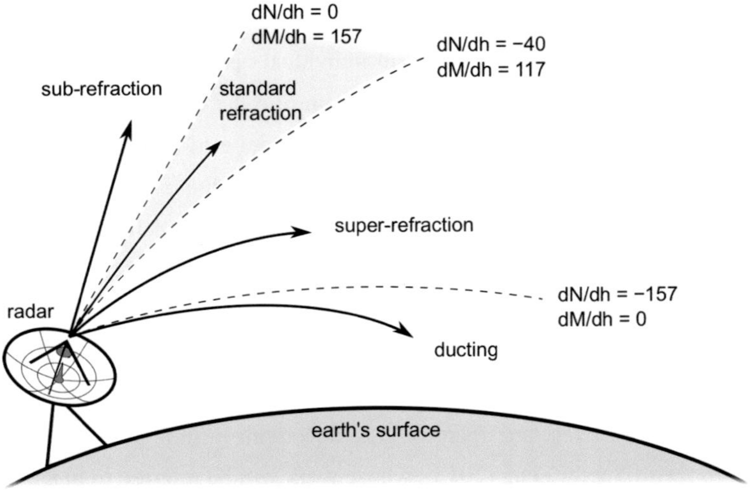

Figure 5.2: Possible cases of refraction of the radar beam due to present atmospheric conditions according to the vertical gradient of the refractivity N or the modified refractive index $M = 10^6 h/R_E + N$ respectively (in analogy to Turton et al., 1988).

In air the refractivity is not a constant but depends on the present conditions as temperature T, pressure p and vapour pressure e according to the following relation (Bean and Dutton, 1966):

$$N = 77.6\frac{p}{T} - 5.6\frac{e}{T} + 3.75 \times 10^5 \frac{e}{T^2}$$
$$N \approx \frac{77.6}{T}\left(p + 4.81 \times 10^3 \frac{e}{T}\right)$$

For example, based on the International Standard Atmosphere with the surface values for $T = 288.15\,\mathrm{K}$ (that equals 15°C) as well as $p = 1013.25\,\mathrm{hPa}$ and assuming a relative humidity of 35%, which means the vapour pressure $e = 6.0\,\mathrm{hPa}$, the resulting refractivity is 300. At saturation (100% relative humidity) with $e = 17.1$ hPa and same temperature and pressure conditions the refractivity would be 350.

It is generally known that the atmospheric parameters vary at most in vertical direction. The pressure as well as the vapour pressure decrease rapidly (exponentially) with altitude. The temperature, however, decreases gradually (linearly) or sometimes even increases with height. Thus, the vertical gradient of the refractivity is negative while the horizontal variations can be neglected. The result is that the upper wave packets of the radar pulse propagate faster than the underlying parts, which causes a bending of the radar beam towards the earth's surface. This effect is called atmospheric refraction. Under normal conditions the radiation follows standard refraction where the curve radius of the bended beam is still smaller than the earth's curve radius. In certain cases the atmospheric circumstances strongly differ from the normal case and consequently lead to a stronger deviation from the standard refraction, the result is sub-refraction, super-refraction or ducting (see Figure 5.2). The latter causes the beam to hit the earth's surface. On rare occasions the radar pulse even can be trapped inside atmospheric layers leading to a wave-like propagation.

The knowledge of the path of the propagating radar pulse is important for the exact localisation of the received reflectivity signal. Concering spatial variations of the refractive index only a vertical change is assumed. Therefore it is sufficient to know the height h_i above MSL (mean sea level) at any range point r_i and arc distance s_i, respectively, along the trajectory in terms of the local incoming elevation angle θ_i. Since the horizontal gradient of the refractive index is negligible the azimuth angle ϕ remains constant along the main axis. The radar forward operator is supplied with three possible configurations how to calculate h (Zeng, 2013):

1. Equivalent earth model
   In the literature (e.g. Doviak and Zrnic, 1993) a simple approximation can be often found based on a standard atmosphere with steady and representative parameters. In this case the vertical gradient of the refractive index is $dn/dh = -4 \times 10^{-8}\,\mathrm{m}^{-1} = const.$ It can be shown that multiplying the earth radius R_E by the factor 4/3, the radar beam would propagate along a curve parallel to the earth's surface if an equivalent 4/3-earth radius is considered. This method is time independent and has the advantage of

computational simplicity. In case of standard refractivity it has adequate accuracy but it can lead to significant errors in the localisation of the radar pulse if the present atmospheric conditions show a strong variation from the assumption which cause anomalous propagation. This error is particularly large in case of ducting.

2. Total reflection
   According to Snell's law for a continuous spherically stratified medium

   $$n(h)\,(R_E+h)\cos\theta_i = const \quad \text{with} \quad \tan\theta_i = \frac{R_E}{R_E+h}\frac{dh}{ds},$$

   h can be calculated by discretisation of the refractive index in constant increments and adopting the criteria of total reflection at each transition of n_i to n_{i+1}. Thus, the path consists of discrete straight ray parts. This method knows the current atmospheric state and is able to capture anomalous propagation but at the expense of computational cost because of its time dependence. Furthermore there is an ambiguity of the correct sign of the incoming elevation θ_i. This problem can be avoided by applying an additional ad hoc criterion (Zeng, 2013).

3. Second-order ordinary differential equation
   The exact description of the problem is given by the Euler-Lagrange-equation (Hartree et al., 1946):

   $$\frac{d^2h}{ds^2} - \left(\frac{2}{R_E+h} + \frac{1}{n}\frac{dn}{dh}\right)\left(\frac{dh}{ds}\right)^2 - \left(\frac{1}{R_E+h} + \frac{1}{n}\frac{dn}{dh}\right)\left(\frac{R_E+h}{R_E}\right)^2 = 0.$$

   This equation can also be formulated as a function of $h = h(r)$ instead of $h(s)$. Based on the altitude and elevation of the antenna this is an initial value problem which can also be solved by discretisation. The case of total reflection is already described by the equation and there is no sign ambiguity. This method has the same advantages as the second method and is more robust at the same time.

A more detailed description of all three methods and some sensitivity experiments can be found in the work of Zeng (2013) and Zeng et al. (2013), respectively.

Beam Weighting Function

As already mentioned in Section 4.1 the radar pulse is not concentrated in one single ray but broadens with distance (cf. Figure 4.1 on page 26). This problem mainly comes up at ranges further away from the radar where the resolution is lowered and the received signal becomes more and more blurred over an increased volume. To deal with this effect the beam weighting function (Blahak, 2008) is introduced, a volume averaging over the radar pulse. Therefore several auxiliary radar axes are calculated that provide additional finite integration points in azimuthal and elevational direction. The number of axes (points) can be chosen arbitrary. The numerical summation over those integration points is done via the Gauss-Legendre quadrature.

By using auxiliary axes to capture the broadening of the radar beam a second problem can be treated at the same time to calculate the received radar observables. A single ray would totally be blocked if hitting a fixed obstacle or the "ground",[9] respectively. However, because of the extent of the radar beam only parts of the pulse are shadowed while other parts could still "see through" the obstacle. This is true as long as at least one auxiliary axis can pass the object. Sensitivity experiments to find the optimal number of axes per radar beam are described in the work of Zeng (2013).

5.3 Simulation of the Radial Velocity

The radial velocity v_r is calculated from the three-dimensional wind components u, v and w of the model variables. After the propagation of the radar beam has been simulated, the model values are interpolated trilinearly on the

[9] See the description of ground clutter in section 4.3 for the definition of the term "ground" in this case.

coordinate points of the radar pulse. In the next step the radial velocity simply can be derived from the wind components given in terms of the radial distance r, azimuth θ and elevation ϕ. Järvinen et al. (2009) and Lindskog et al. (2004) proposed the following equation, only considering the horizontal wind field while the vertical wind component is neglected at first:

$$v_h = u\sin\theta + v\cos\theta \quad \rightarrow \quad v_r = v_h\cos(\phi + \alpha).$$

The sign of v_h is positive in agreement with the convention described in paragraph 4.2.2.

For the use within the radar forward operator all wind components should be taken into account. So, the equation is extended by the vertical term analogously

$$v_r = v_h\cos(\phi + \alpha) + w\sin(\phi + \alpha).$$

The angle α results from the curvature of the earth's surface. In case of the equivalent earth model (4/3-earth) this argument can be approximated after Järvinen et al. (2009) by

$$\alpha = \arctan\left(\frac{r\cos(\phi)}{r\sin(\phi) + \frac{4}{3}R_E + h_0}\right),$$

where R_E is the radius of the earth and h_0 is the height of the antenna above mean sea level. In case of a variable refractive index of the atmosphere α has to be estimated directly when simulating the propagation of the radar beam.

There are two further options that can be considered in the modular radar forward operator when simulating the radial velocity. First, when regarding the vertical wind component, the average terminal fall speed of the hydrometeors have an impact on it and can be taken into account for additional accuracy. That becomes significant especially at high elevations. Second is the weighting by the reflectivity values. Some more detailed remarks on the simulation of the radial velocity can also be found in Zeng (2013).

Table 5.1: List of all considered cases of hydrometeors according to their compound of state of matter. Here, "dry" means completely frozen and "wet" means partially melted.

	state of matter	example
water	pure liquid water	cloud water, raindrops
ice	pure frozen water	cloud ice, dry hail
ice/air	mixture of frozen water and air	dry snow, dry graupel
ice/water	mixture of frozen and liquid water	wet hail
ice/air/water	mixture of frozen water, air and liquid water	wet snow, wet graupel

5.4 Simulation of the Radar Reflectivity

5.4.1 Refractive index

As already introduced in paragraph 4.2.1, the reflectivity is a function of the backscattering cross-section of the scatterer which in turn mainly depends on its refractive index. So, first of all the refractive index of every type of hydrometeor has to be calculated, which is a measure for the dielectric properties of the media. The refractive index n was introduced in the previous section as a factor that determines the propagation velocity of electromagnetic radiation. However, generally the refractive index m is a complex variable denoted as

$$m = n - ik,$$

where the real part n is the part of the refractive index that conforms the definition $n = c/c_0$. The imaginary part k is referred to as the absorption coefficient and would be zero in a lossless dielectric, which is nearly fulfilled for atmospheric air. Hence, only the real part of the refractive index is used in the calculation of the propagation of the radar beam (cf. section 5.2).

When calculating the scattering of hydrometeors (liquid and frozen water), the complete complex form has to be taken into account for any given classification, furthermore the hydrometeor's state of matter plays a decisive role. It should be borne in mind that besides pure liquid and frozen water there are also mixtures of states as, e.g. melting particles (cf. Table 5.1) and it becomes even more complicated considering to define a reliable refraction index (or scattering cross-section). In this case an additional parameter has to be calculated, the degree of melting, which is not predicted by the microphysical scheme in the COSMO model and has to be estimated from other available information (Blahak, 2009). This dimensionless, temperature dependent quantity determines the fraction of liquid water within the particle and is "0/1" at a predefined minimum/maximum temperature.

Table 5.1 shows all possible cases with the corresponding examples of hydrometeors that are included in the COSMO model so far. This fact poses a serious challenge in calculating the refractive index. The task is to consider the mixture as a homogeneous medium with an effective refractive index[10] m_{eff}. There are several different formulations of calculating the refractive index of both pure and mixed states depending on the composition of the material (Blahak, 2009). Additionally the refractive index is a function of temperature T and wavelength λ. For the radar forward operator the following formulations/parameterisations of m_{eff} are implemented, which the user can choose (Blahak, 2009):

- for pure liquid water:
  - Ray (1972): valid for $-10°\text{C} < T < 30°\text{C}$ and $0.001\,\text{m} < \lambda < 1.0\text{m}$.
  - Liebe et al. (1991): valid for $-3°\text{C} < T < 30°\text{C}$ and $0.0003\,\text{m} < \lambda < 0.3\text{m}$.
- for pure frozen water:
  - Ray (1972): valid for $-20°\text{C} < T < 0°\text{C}$ and $0.001\,\text{m} < \lambda < 10^7\text{m}$.
  - Mätzler (1998): valid for $-60°\text{C} < T < 0°\text{C}$ and $10^{-9}\,\text{m} < \lambda < 8.6\text{m}$.

[10]The method is commonly called effective medium approximation.

- Warren (1984): valid for $-250°C < T < 0°C$ and $0.0001\,m < \lambda < 10^7 m$.

- for mixed states (ice/air, ice/water, ice/air/water):
  - Oguchi (1983): homogeneous mixture of inclusions with shape parameter u ($u = 2$ in case of spherical inclusions) in a isotropic background material.
  - Maxwell-Garnett (1904): spherical or spheriodical inclusions with random size in a isotropic background material, leading to 15 different possibilities of defining m_{eff} of a three-phase mixture (ice/air/water).
  - Bruggemann (1935): homogeneous mixture of small spherical grains of different material equally distributed in a enclosing sphere combining a dense package.

For further calculations only the formulation of Ray for liquid water and Mätzler for frozen water is used.

5.4.2 Rayleigh and Mie Theories

In paragraph 4.2.1 the radar reflectivity was defined by equation (4.3) on page 30 by

$$\eta = \int_0^\infty dD\, \sigma_b(D) N(D)\,. \qquad \text{(4.3 revisited)}$$

The particle size distribution $N(D)$ of each hydrometeor type is already defined by the COSMO model (see section 3.3) and will be adopted unmodified. The backscattering cross-section σ_b is a new, as yet not described quantity that indicates the part of energy which is scattered back towards the direction of incidence. It corresponds to a fictional area of the target which is "seen" by the radar and is a function of the particle's diameter D (and shape), its refractive index m and the wavelength λ of the interacting radiation, i.e. $\sigma_b = \sigma_b(D, m, \lambda)$. After the refractive index has been calculated, the backscattering cross-section is determined in the next step.

Gustav Mie (1908) found a solution for the cross-section of dielectric homogeneous spherical particles directly from Maxwell's equations

$$\sigma_b = \frac{\lambda^2}{4\pi} \left| \sum_{n=1}^{\infty} (-1)^n (2n+1)(a_n - b_n) \right|^2 , \tag{5.1}$$

the Mie coefficients a_n and b_n are spherical Bessel functions depending on m and a parameter $\alpha = \pi D/\lambda$. This theory provides the most accurate description of the problem but comes at a high computational cost. There are efficient formulations to calculate the backscattering cross-section according to Mie's theory (e.g. from Bohren and Huffman, 1983). Furthermore, different algorithms for two-layered particles[11] are implemented in the radar forward operator, one directly based on Kerker (1969) and another given by Bohren and Huffman (1983). For the further simulation only the latter is used, though Kerker gives a more general but also more complicated solution which is omitted here, for brevity.

An alternative to the expensive but exact Mie solution is an approximation ascribed to Rayleigh

$$\sigma_b = \frac{\pi^5}{\lambda^4} |K|^2 D^6 \quad \text{with} \quad K = \frac{m^2 - 1}{m^2 + 2} . \tag{5.2}$$

This is a good simplification and accurate enough in most cases. The dielectric factor $|K|^2$ depends only on the refractive index and is about 0.93/0.18 for liquid/frozen water. However, the Rayleigh approximation has a significant limitation. It is only valid for small particles compared to the wavelength, that means in cases when $D \ll \lambda$ or $\alpha \ll 1$, respectively. The reason for this can be seen in Figure 5.3. For greater particles the approach becomes inaccurate and the size of the particles related to the backscattered energy will be

[11] only for (melting) snow

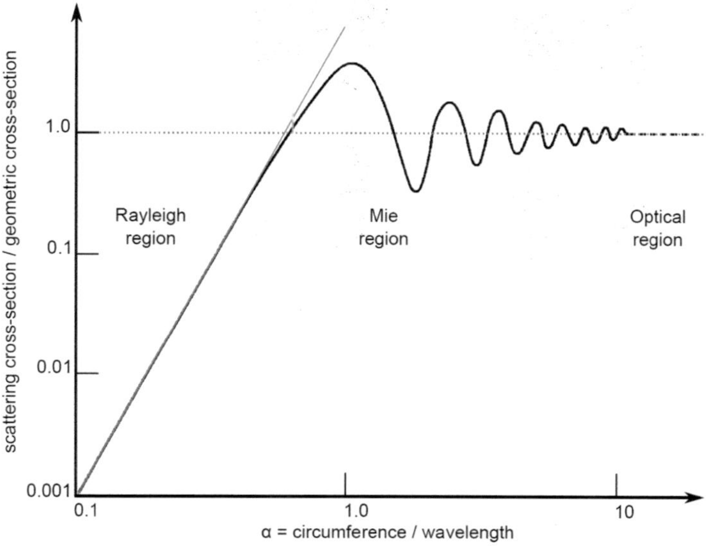

Figure 5.3: The normalised cross-sectional area of a perfectly conducting metal sphere with constant diameter D related to its optical area as a function of the parameter $\alpha = \pi D/\lambda$ depending on the wavelength λ (in analogy to Skolnik, 1980). This demonstrates the range of validity of the Rayleigh approximation indicated by the solid tangent straight line.

overestimated. At most radar wavelengths the Rayleigh approximation can be used for cloud particles, rain and small frozen hydrometeors. But with larger particles like graupel or hail the exact Mie solution should be considered, as the backscattering cross-section shows an oscillatory behaviour (as can also be seen in Figure 5.3) because of interference effects. And for even bigger objects the scattering cross-section of the target approaches their geometric cross-section.

It should be noted that both the Mie and the Rayleigh theory are valid for spherical particles only. However – though most hydrometeors deviate from a spherical form – the COSMO model only knows the diameter of a particle considering it as a sphere and hence any asphericity will be neglected. The task of considering the particle's actual shape remains for future research.

Inserting equation 5.2 in equation 4.3, the radar reflectivity according to the Rayleigh approximation becomes

$$\eta = \frac{\pi^5}{\lambda^4}|K|^2 \int_0^\infty dD D^6 N(D) \quad \text{and} \quad Z = \int_0^\infty dD D^6 N(D) \tag{5.3}$$

$$\Rightarrow \quad Z = \frac{\lambda^4}{\pi^5 |K|^2}\eta \,. \tag{5.4}$$

Here, the radar reflectivity factor Z emerges again, as already foreshadowed in paragraph 4.2.1. This new quantity was originally introduced in order to get a parameter that is independent of the radar's wavelength, which is only true if the Rayleigh approximation is valid. Using this approach Z is related to the sixth moment of the particle size distribution. From a historical point of view and for reasons of clarity the unit of Z is given im $\mathrm{mm}^6/\mathrm{m}^3$. The general form can be obtained by inserting equation (4.3) for η in equation (5.4) for Z. Hence, the radar reflectivity factor is defined as

$$Z = \frac{\pi^5}{\lambda^4}|K|^2 \int_0^\infty dD\, \sigma_b(D) N(D) \,.$$

In paragraph 4.2.1 also the equivalent radar reflectivity factor was introduced, which is derived by replacing the general dielectric factor $|K|^2$ by the specific (nearly) constant value for liquid water $|K_w|^2 = 0.93$

$$Z_e = \frac{\pi^5}{\lambda^4}|K_w|^2 \int_0^\infty dD\, \sigma_b(D) N(D) \,.$$

This convention has to be done for technical reasons. Since the radar software does not know the actual state of matter of the scatterers, small spherical liquid water particles are assumed. All further considerations of the reflectivity refer to this quantity (unless otherwise noted).

It should be mentioned that the refractive index, the backscattering cross-section and the particle size distribution are describing only one type of hydrometeors, respectively, and technically speaking all previous expression only describe one type of hydrometeor. Consequently, an index denoting the particle type has to be added consistently to all parameters, e.g. K_r, $\sigma_{b,r}$,etc. for rain. In order to get the total reflectivity a sum over all categories has to be taken

$$Z_e = \sum_j Z_{e,j} \quad \text{with} \quad j = \{\text{Classes of hydrometeors}\}\,.$$

However, the index is previously omitted and will be in the following also to improve readability[12].

So far, the radar reflectivity can be calculated at every grid point of the COSMO model but no effects of the propagation of the radar beam (refraction, broadening of the radar beam) are taken into account. Before going further on that, the influence of the attenuation has to be described.

5.4.3 Attenuation

Attenuation of the radiation can be described by Lambert-Beer's law. Accordingly, extinction results in an exponential decay of the radiation (amplitude, power) during the transmission through a medium

$$\ell^{-2}(r, \phi_p, \theta_p) = \exp\left(-2\int_0^r dr' \Lambda(r', \phi_p, \theta_p)\right)\,.$$

The quantity ℓ^{-2} already appeared in the radar equation (4.2) in section 4.1 and is called the two-way attenuation coefficient. The extinction coefficient Λ is definded analogically to equation (4.3) of the radar reflectivity η, by replacing the backscattering cross-section σ_b with the extinction cross-section σ_{ext}

$$\Lambda = \int_0^\infty dD \sigma_{ext}(D) N(D)\,. \tag{5.5}$$

This term has to be integrated along the path of propagation, though it should be remembered that the attenuation takes place twofold – on the way to the target

[12] The same consideration also applies to the attenuation.

and on the same way back to the radar. Here and in the following, only the extinction by hydrometeors is considered.

Electromagnetic radiation can be attenuated in two ways by interacting with the hydrometeors. A part of the energy is absorbed by the medium and transformed into heat. Another part is scattered out of the main direction of the beam. But only the backscattered part is measured by the radar and only the transmitted part is available for further detection along the path while all energy which is scattered sidewards is lost for the measurement. So, the total extinction cross-section is composed of the absorption term and the losses due to scattering:

$$\sigma_{ext} = \sigma_{abs} + \sigma_{sca}\,.$$

The exact formulation of the extinction cross-section is again given by the Mie theory as

$$\sigma_{ext} = \frac{\lambda^2}{2\pi}\left(-\Re\sum_{n=1}^{\infty}(2n+1)(a_n+b_n)\right),$$

with the same parameters a_n and b_n as occurring in equation (5.1) for the backscattering cross-section. Also there is a more general solution for a_n and b_n for two-layered spheres from Kerker (1969) and efficient code from Bohren and Huffman (1983).

And again there is a small-size approximation similar to the Rayleigh theory that was used in equation (5.2). In case of extinction this is

$$\sigma_{ext} = \frac{\pi^2}{\lambda}\Im(-K)D^3 + \frac{2\,\pi^5}{3\,\lambda^4}|K|^2D^6\,, \tag{5.6}$$

with the first term describing the absorption and the second term describing the attenuation due to scattering. However, this approach is less accurate than for the backscattering cross-section and should only be used for very small particles. For this reason it is forgoed to consider attenuation when using the Rayleigh theory in the radar operator.

An exception is the attenuation by clouds. In this case the Rayleigh approximation is sufficient. Moreover, the scattering term can be neglected and only the extinction by absorption is considered. Inserting the first term of equation (5.6) in equation (5.5) the extinction coefficient is (Doviak and Zrnic, 1993):

$$\Lambda_c = \frac{\pi^2}{\lambda} \Im(-K) \int_0^\infty dDN(D)D^3 \,.$$

The integral is related to the third moment of the size distribution. In section 3.2 this was defined as the liquid water content L described by equation (3.2) on page 17

$$L = \frac{\pi \rho_w}{6} \int_0^\infty dDN(D)D^3 \,.$$

Now, using the liquid water content L the attenuation by clouds can be expressed as:

$$\Lambda_c = \frac{6\pi^2 L}{\rho_w \lambda} \Im(-K) \,.$$

The extinction coefficient Λ is calculated analogously to the equivalent radar reflectivity factor Z_e on the model grid. However, it cannot be applied to Z_e before knowing the propagation path. So, the next step is to calculate the two-way attenuation coefficient ℓ^{-2} by integrating Λ along the path after the propagation of the radar beam is calculated (regardless which method is used). Now, the attenuated reflectivity can be expressed as the product of both $Z_{ext} = Z_e \ell^{-2}$. Finally the broadening of the radar pulse can be included using the beam weighting function. The resulting mean is given by:

$$\overline{Z}_{ext}(\vec{r}) = \frac{\iiint\limits_V dV \; Z_e(r,\phi_p,\theta_p)\ell^{-2}(r,\phi_p,\theta_p)\frac{f^4(\phi_p,\theta_p)}{r^4}}{\iiint\limits_V dV \; \frac{f^4(\phi_p,\theta_p)}{r^4}} \,,$$

with $dV = dV(r,\phi_p,\theta_p) = r^2 \cos\theta_p d\theta_p d\phi_p dr$.

5.5 Lookup Tables

As already stated, the exact formulation of the cross-section is given by the Mie theory (and its generalised version for two-layered particles) and hence the best possible method to calculate reflectivity and extinction. In contrast, using the Rayleigh approximation, attenuation is not even included within the radar forward operator. Furthermore, the Mie theory is not only the best choice in case of larger particles but also when melting particles occur due to the additional consideration of one- or two-layered spheres. However, this accuracy has its price of high computational complexity. Without having to waive either good quality or efficiency a compromise has to be found. One solution are lookup tables, lists of sampling points precalculated once (in the first time step) and then used in further computational runs. The actual values will be interpolated from those points. Since this process is faster than the complex calculation of the Mie coefficients this method reduces the overall computation time.

In the case of the calculation of reflectivity and attenuation again every hydrometeor type to which the Mie theory is applied has to be considered separately and hence has to have its own array of sampling points. So, in case of considering graupel this means creating a lookup table for the reflectivity values for rain as well as for completely frozen ("dry" snow and graupel) particles and partly melted ("wet" snow and graupel) particles, respectively, as all these components contribute to graupel formation. The same is to be done also for the attenuation values. Accordingly, this means ten distinct lookup tables have to be created in total.

With regard to rain and other frozen (not-melting) particles the reflectivity and the extinction are functions of their corresponding cross-sections and the particle size distribution of the respective type of hydrometeor, see equation (4.3) and equation (5.5), respectively. In turn these two parameters are depending on the particle refractive index m and the total mass fraction $q_{x=\{r,s,g\}}$. The refractive index can be reduced to a function of the temperature T for liquid drops and a function of T and q_x for frozen particles. Thus, depending on the number of sampling points the lookup table will be an array of $N_T \times N_{q_x}$ entries. With regard to melting snow and graupel, a third coordinate in the lookup table has to be added that determines the degree of melting, because in this case m also depends on the degree of melting. This can be be indicated by the maximum temperature T_{max} at which the particle will be totally melted. So, in this case the lookup table is an array of $N_T \times N_{q_x} \times N_{T_{max}}$ entries. Denoting T_{min} as the temperature at which the frozen particle starts to melt, the whole melting process takes place in the temperature range between T_{min} and T_{max}.

The initial as well as the final value of each variable and the number of interim values can be set arbitrarily. For the radar forward operator the following configurations for the lookup tables were currently chosen for each category:

rain:

- $T = [-40, 40]$°C with $N_T = 20$
- $q_r = [-7, -1]$ kg/kg with $N_{q_r} = 30$

dry snow:

- $T = [-80, T_{min}]$°C with $N_T = 40$, $T_{min} = 0$°C
- $q_s = [-6, -2]$ kg/kg with $N_{q_s} = 15$

dry graupel:

- $T = [-80, T_{min}]$°C with $N_T = 20$, $T_{min} = -10$°C
- $q_g = [-6, -1]$ kg/kg with $N_{q_g} = 30$

wet snow:

- $T = [T_{min}, 10]$°C with $N_T = 50$, $T_{min} = 0$°C
- $q_s = [-6, -2]$ kg/kg with $N_{q_s} = 15$
- $T_{max} = [3, 10]$°C with $N_{T_{max}} = 14$

wet graupel:

- $T = [T_{min}, 15]°C$ with $N_T = 50$, $T_{min} = -10°C$
- $q_s = [-6, -1.5]$ kg/kg with $N_{q_g} = 30$
- $T_{max} = [3, 15]°C$ with $N_{T_{max}} = 24$

It should be noted that in case of snow (dry and wet) one problem arises for the calculation of the lookup tables since the parameter $N_{0,s}(T)$ in the particle size distribution for snow is also temperature dependent (cf. section 3.3). Therefore this parameter is calculated separately before each single entry in the table.

After all lookup tables have been created in the precalculation part (first time step) of the model the interpolation of the actual values is carried out bilinealy and trilinearly, respectively, in the operational runs. This is done in three steps: 1. The actually given values of q_x, T and, where necessary, T_{max} are truncated to the range of the lookup table. 2. The indices of the neighbouring regular values in the table are computed. 3. The interpolation between these points is carried out. Once the lookup table has been created the calculation of the actual values of reflectivity and attenuation by interpolation is a much faster process and saves a lot of computational time. It turns out to be even faster than the simpler calculation using the Rayleigh approximation.

5.6 Case Studies

5.6.1 Idealised Case Studies

In the previous sections the radar forward operator has been described in great detail with the main focus on the simulation of the radar reflectivity and the attenuation of radiation during the interaction with hydrometeors. In this section the operator's reliability shall be demonstrated. But before it can be applied to real synoptical situations some test cases have to be done first. The first simulations were performed in an idealised environment with initial conditions along the lines of Weisman and Klemp (1982) using a convective system which was triggered by three simultaneously released warm bubbles. These thermal perturbations injected, in a apart from that horizontally homogeneous atmosphere (Figure 5.4, left upper plot), lead to the development of a typical convective storm structure which results in a lot of precipitation. Therefore, this test case is well suited for the simulation of the radar measurement of hydrometeors.

The radar reflectivity without attenuation and averaging over the radar pulse volume is available in both, the coordinate system of the model and after interpolation on the radar beam given in spherical coordinates. Initially, it is calculated on the COSMO model grid and hence independently of any radar characteristics but merely dependent on the prognostic model variables. This means in the first calculation step no radar station is needed. So for a start, Figure 5.4 shows the calculated reflectivity given on the model grid resulting from the idealised test case described above after four hours of simulation. The graphics are pseudo-three-dimensional illustrations in a top view together with the side views. In each view the projection of the maximum reflectivity is shown. This is a simple form to visualize a three-dimensional measurement with focus on high reflectivity values.

First of all it should be mentioned that the values cover a wide range of magnitude and thus it is more convenient and usual practise in radar meteorology to convert these values from linear units to logarithmic units, using the dB-scale referred to Z that is defined as

$$\xi_e[dBZ] = 10\log_{10} Z_e[\text{mm}^6\text{m}^{-3}].$$

Secondly, it can be seen from the figure that periodic boundary conditions are used, meaning that model values on one margin simultaneously provide the boundary data on the opposite margin. So, the developing precipitation that moves out of the model domain comes in again on the other side. Figure 5.4 shows the temporal development of the convective cell with plausible resulting values of the radar reflectivity.

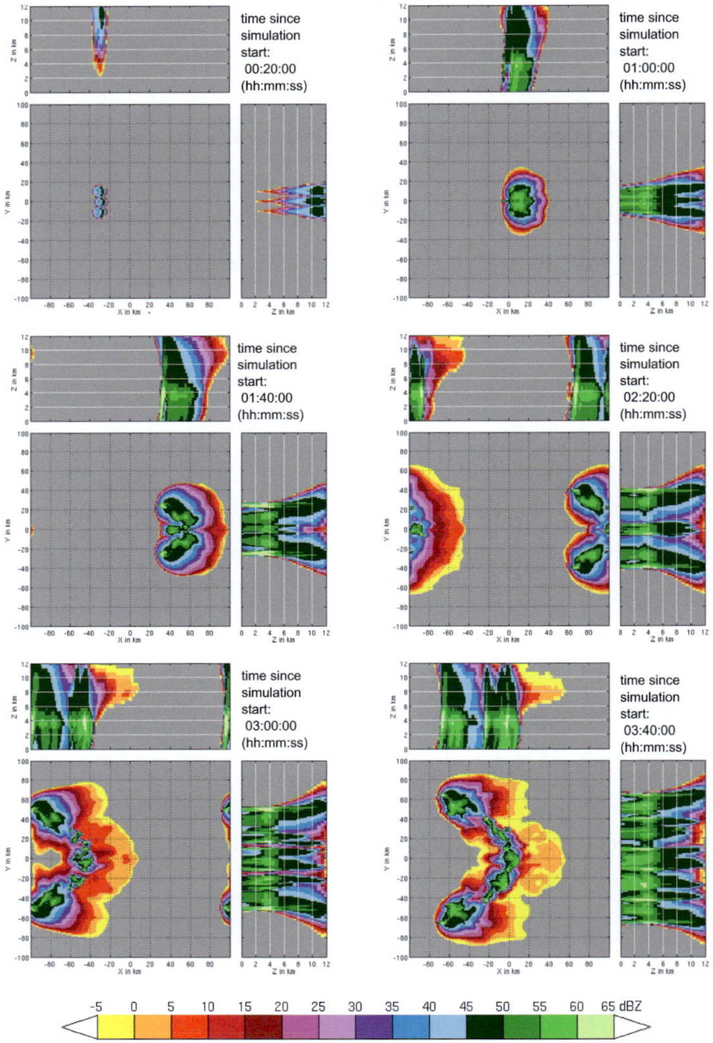

Figure 5.4: Simulated radar reflectivity (in dBZ, see colour bar) of an idealised test case in time intervals of 40 min, starting at the upper left plate 20 min after injecting three warm bubbles, and ending after 3 h 40 min (lower right plate). The values are plotted on the model grid with the projection of the maximum reflectivity in top and side view. Here, no radar station is considered.

In the next step the reflectivity values are interpolated by the radar forward operator to the centres of an assumed radar pulse volume according to their range, azimuth and elevation. At first, only one radar station was taken into account having a resolution of 130 range bins of 1 km in radial direction, 20 elevations and 360 azimuths in steps of 1°. The results of the same test case is shown in Figure 5.5 and Figure 5.6, respectively, using a display of a complete single elevation scan as an another alternative of visualisation, a so-called plan position indicator (ppi). This is a simple way to illustrate the radar reflectivity on a radar beam, since no transformations have to be made, when the data are already existing on the radial grid. Unfortunately, this is only a very limited view and in order to get a complete picture of the three-dimensional distribution of the precipitation a ppi-plot of each of the 20 elevations would be necessary. To demonstrate this, Figure 5.5 represents a selection of ppi-plots for different elevation angles at a specific point in time (after 2 h of simulation). But in most cases it is sufficient to look at just one elevation, as in Figure 5.6 showing the same temporal development as Figure 5.4. In this case the ppi of the 1.5°-elevation is chosen. Note that when comparing the plots of the model values of the reflectivity (cf. Figure 5.4) with the ppi-plots in the latter graphics the centre is the position of the radar station, which is different to the centre of the model domain.

To complete this discussion, the simulated radial velocity (using the same test case) can be seen in Figure 5.7 as ppi with the same configurations as employed in Figure 5.6. The radial velocity given by the radar forward operator was plotted unmodified, which means including aliasing (the description was given in paragraph 4.2.2). This is clearly visible in the sharp transitions from blue (-20 m/s) to red (20 m/s) wind fields occurring in the last two panels of Figure 5.7, where velocities exceeding the maximum range are folded. All in all, a successfully simulation of the radial velocity could be shown.

The presented test case of Weisman and Klemp (1982) with the use of three warm bubbles is the basic principle for the further idealised case studies. Now,

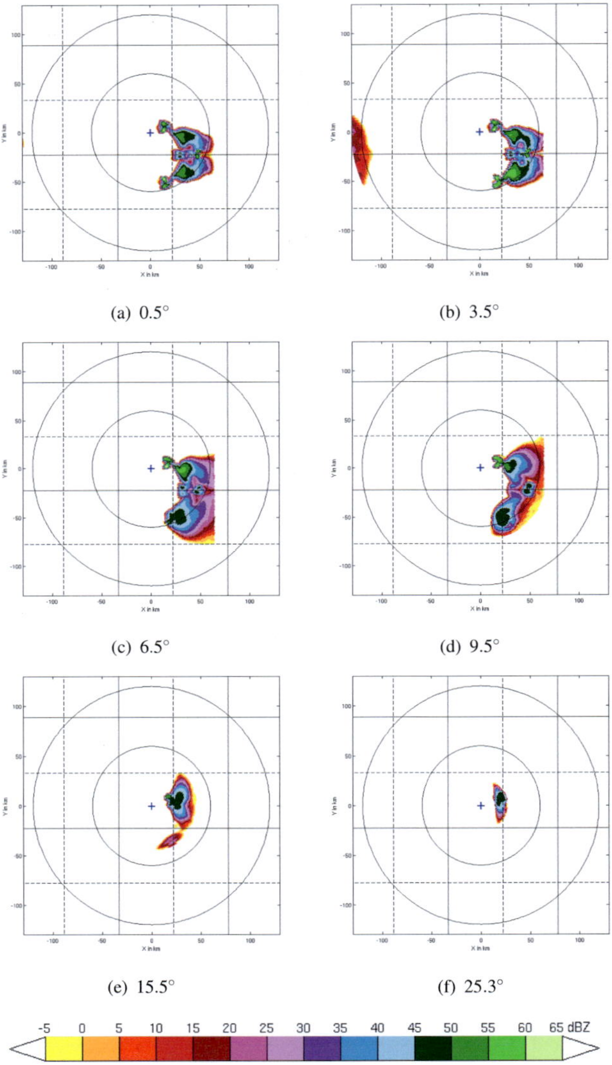

(a) 0.5°

(b) 3.5°

(c) 6.5°

(d) 9.5°

(e) 15.5°

(f) 25.3°

Figure 5.5: Simulated radar reflectivity of the same test case as in Figure 5.4 after 2 h. Now the values are interpolated on the radar beam with the radar station located in the centre of each plot, showing the ppis of different constant elevation scans.

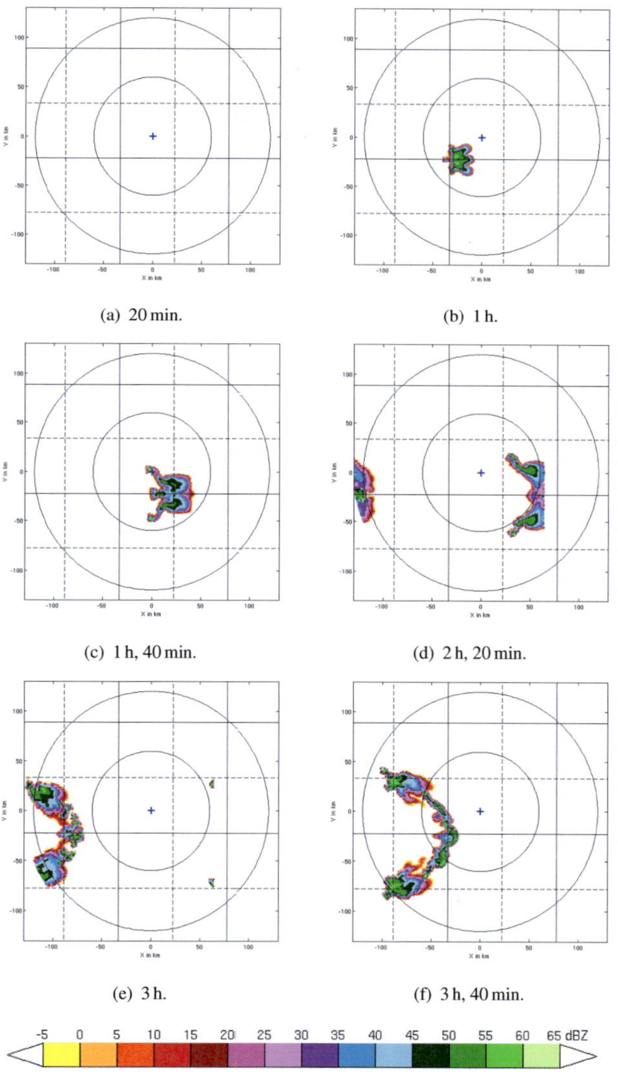

(a) 20 min.

(b) 1 h.

(c) 1 h, 40 min.

(d) 2 h, 20 min.

(e) 3 h.

(f) 3 h, 40 min.

Figure 5.6: As Figure 5.5, but for a constant elevation of 1.5° and different simulation times (same times as in Figure 5.4).

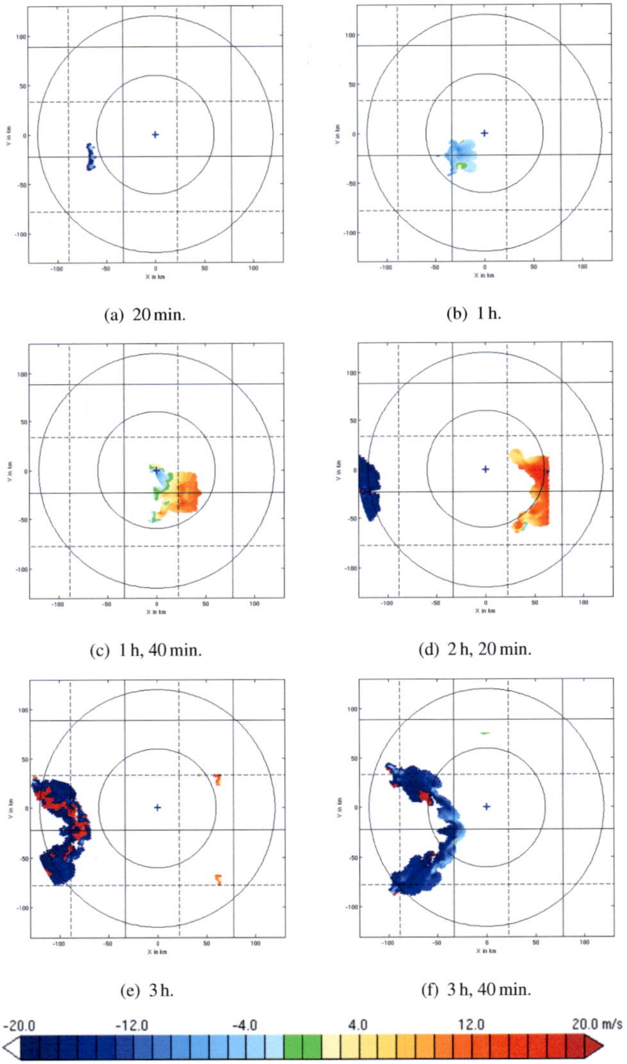

(a) 20 min. (b) 1 h. (c) 1 h, 40 min. (d) 2 h, 20 min. (e) 3 h. (f) 3 h, 40 min.

Figure 5.7: As in Figure 5.6 but for simulated radial velocity (in m/s, see colour bar). In the last two panels aliasing is clearly visible.

various configurations of the radar forward operator are tested for the simulation of the radar reflectivity. The results are shown in Figure 5.8. Two different radar wavelengths are used: $\lambda_1 = 5.5\,\mathrm{cm}$ in the left column and $\lambda_2 = 3.0\,\mathrm{cm}$ in the right column. The results in the first line (a) and (b) have been obtained by applying Rayleigh scattering in combination with the Oguchi approximation of the hydrometeors refractive index (cf. paragraph 5.4.1). Compared with Mie scattering shown in the plots in the second line (c) and (d), it can be seen that the Rayleigh calculation becomes inaccurate at high values of reflectivity. The reason can be the presence of partially melted large graupel. The third line comprising the plots (e) and (f) show the same Mie calculation but additionally applying the attenuation by hydrometeors (cf. paragraph 5.4.3). In this case the strong impact of the extinction especially towards smaller wavelengths is clearly visible. All in all the Figures show the expected results of the simulations. In all cases the propagation of the radar beam according to the simple equivalent earth model is used and no beam broadening is regarded. For further case studies concerning these characteristics see Zeng (2013).

Now finally, having a special look on the attenuation, the two-way attenuation coefficient introduced in paragraph 5.4.3 can also be written in a logarithmic scale (ignoring the integration along the range of the radar beam in this case)

$$k_2[dB] = 10\log_{10}(l^{-2}) = 10\log_{10}(\exp(-2\Lambda)) = -\frac{20\Lambda}{\ln 10}.$$

This can be compared to the unattenuated radar reflectivity on the model grid. The resulting relation can be seen in Figure 5.9(a) first in theory (Blahak, 2004) with different values of the degree of melting f_s for snow and f_g for graupel in comparison with the result of the simulation in Figure 5.9(b). The simulation only contains graupel and rain, because hail is not implemented in the one-moment graupel-scheme, and snow does not occur because the temperature of the test case is too high ($\sim 27°\mathrm{C}$ at the surface). In the plot of the theoretical results, dry particles are omitted because their attenuation effect is negligible (except for dry hail), which can be seen at the little contribution of dry graupel

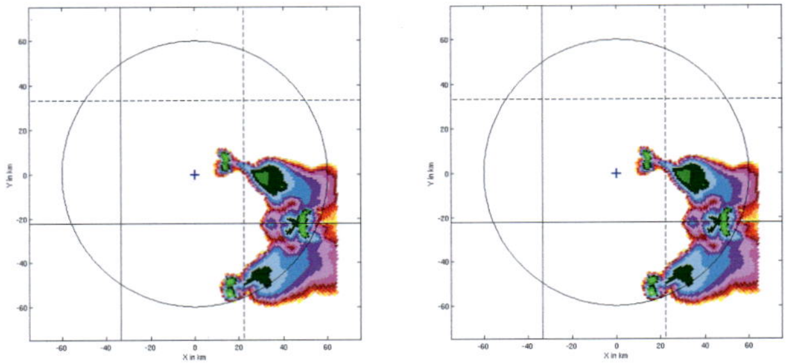

(a) Rayleigh theory, λ_1. (b) Rayleigh theory, λ_2.

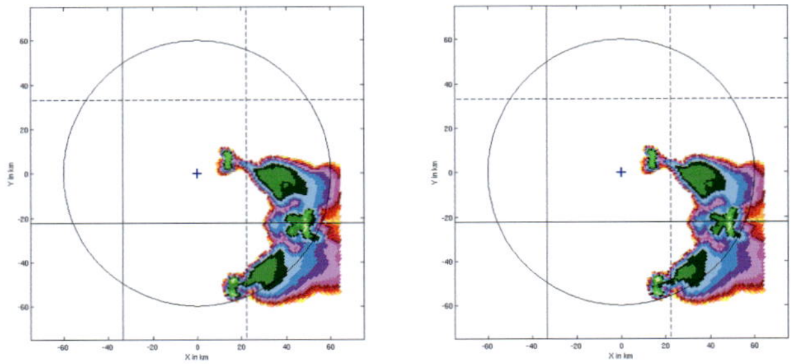

(c) Mie theory, λ_1. (d) Mie theory, λ_2.

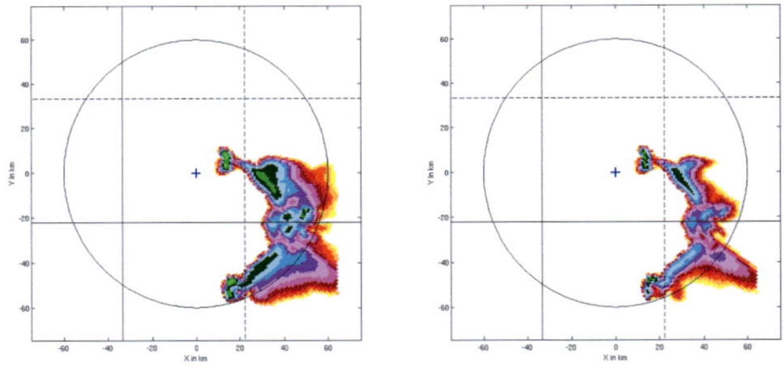

(e) Same as (c) including attenuation. (f) Same as (d) including attenuation.

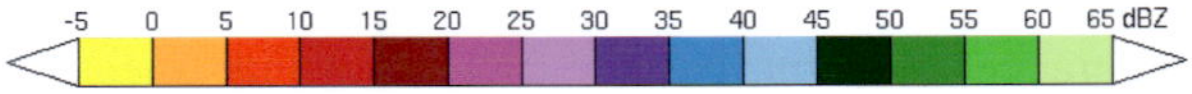

Figure 5.8: Different possibilities (as described in paragraph 5.4.2 and 5.4.3) of simulating the reflectivity (first row: Rayleigh theory, second row: Mie theory without attenuation, third row: Mie theory with attenuation), and for two different radar wavelengths (left column: $\lambda_1 = 5.5\,\text{cm}$, right column: $\lambda_2 = 3.0\,\text{cm}$).

in Figure 5.9(b). But apart from that there is a good agreement for the curves of rain and melting graupel between both pictures which suggests that the simulation of the reflectivity and attenuation is correct.

5.6.2 Real Case Studies

The simulation of the radar reflectivity could be successfully tested using idealised case studies. In the next main step, true meteorological situations were used to verify the reliability of the radar forward operator. And furthermore, not only one radar station was regarded but the complete radar network of the DWD installed throughout Germany. Thus, the calculated values of the reflectivity can be compared to the real measurements. Therefore the domain of simulation is extended to the whole domain of the COSMO-DE model. Figures 5.10 to 5.13 show different meteorological situations with the results of both for the simulated and the measured data. The observations stem from up to 17 radar stations, though occasional technical failures of some stations can be recognised by the forward operator and hence these data are not simulated (as there is no use for it).

The graphics show a pseudo-composite of the lowest ppi-plots – the 1.5° constant elevation scans of each radar station. However, one problem in combining several ppi-plots at different locations arises considering the overlapping areas within the range of two or more radar stations. Since the altitude of one elevation scan is not constant but increases with radial distance, sharp edges at the intersections can occur when different radar data encounter one another at different heights. In order to avoid this problem and to provide smooth transitions in all regions only the reflectivity with the lowest altitude is plotted at locations where more than one radar covers the area. So, the resulting figures show a composite of the values, which are nearest to the earth's surface. This method works well for the simulated data and a smooth composite picture is obtained. But the limitation of this method can be seen in some graphics of the observa-

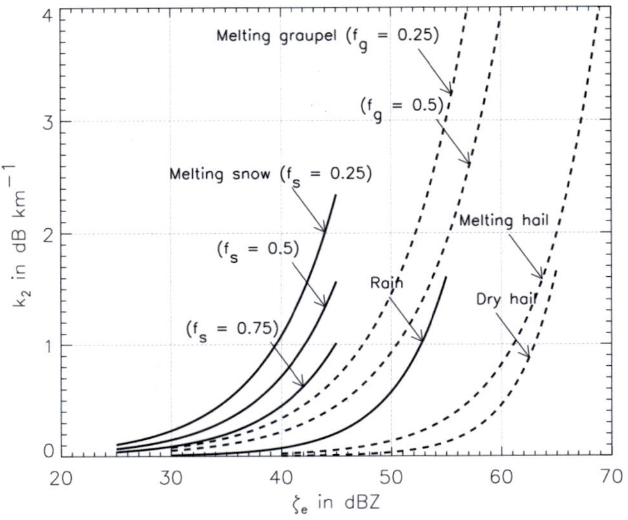

(a) Theory (Blahak, 2004).

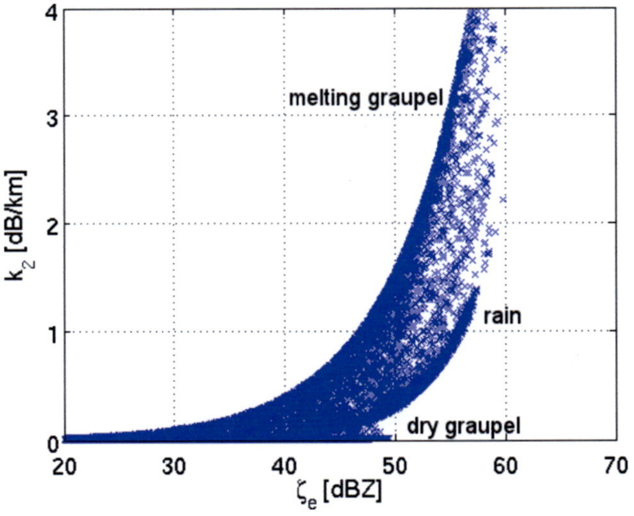

(b) Simulation.

Figure 5.9: Relation between the simulated reflectivity and the two-way attenuation coefficient (lower plate) show good agreement to the theoretical results (upper plate). For details see text.

tional data, which appears as "jumps" of the data in the intersections, mostly occurring at smaller values of the reflectivity.

In the plots of the observational data also some typical errors of the measurement are visible (cf. section 4.3). Second-trip echoes recognisable by the striped structure appear mainly in the Figures 5.12 and 5.13. Clear-air echoes are mainly visible in the vicinity of the radar stations and are only slightly visible. Ground clutter only occurs at one radar station near Munich because of the Alps in the south. The second station that partly covers the region of the Alps is located on the Feldberg and hence high enough (1517 m a.s.l.) not to encounter the mountains. All these errors are detected in a quality check and will be ignored for the operational usage of the data.

The first look at the results shows on the hand the reliability of the radar forward operator in the simulation of radar data. Also the model is able the capture the meteorological situations quite well. On the other hand, it is clearly visible that the model often overestimates the reflectivity at these low heights compared to the measured data. In the next chapter this observation will be examined in more detail.

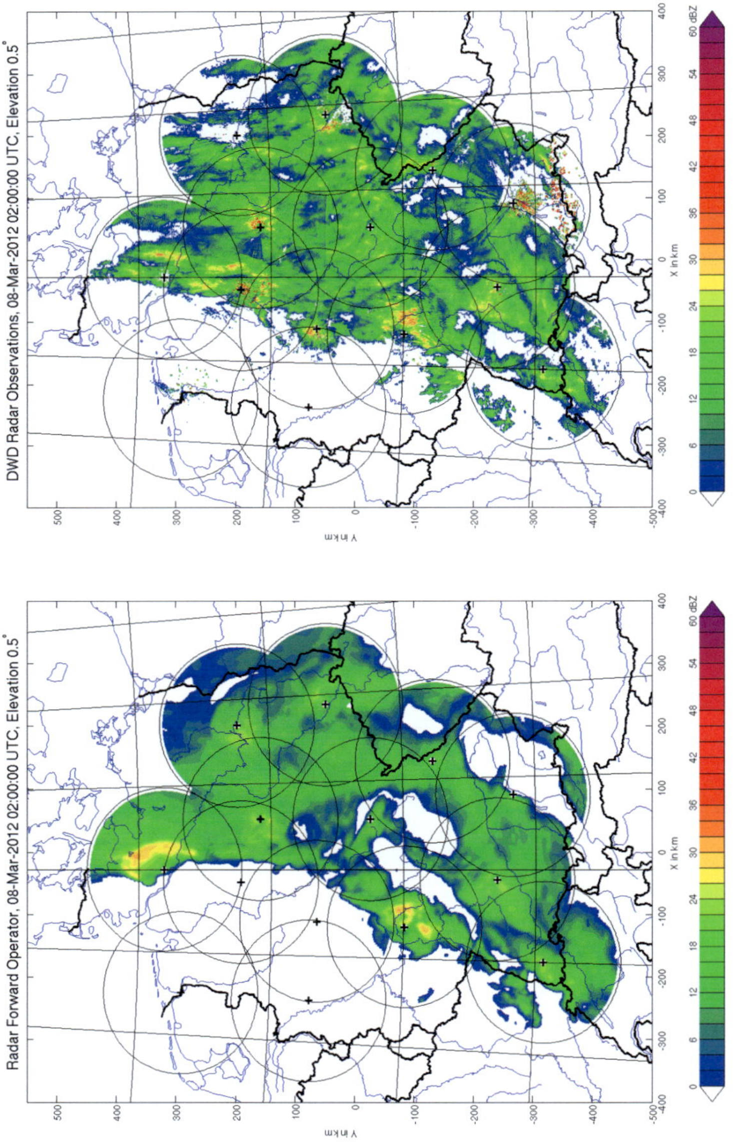

Figure 5.10: Simulation (left) of the radar reflectivity (in dBZ, see color bar) in comparison with observation (right) of a real meteorological case for the region covered by the radar network of DWD: for 08.03.2012, 02 UTC. For plot details see text.

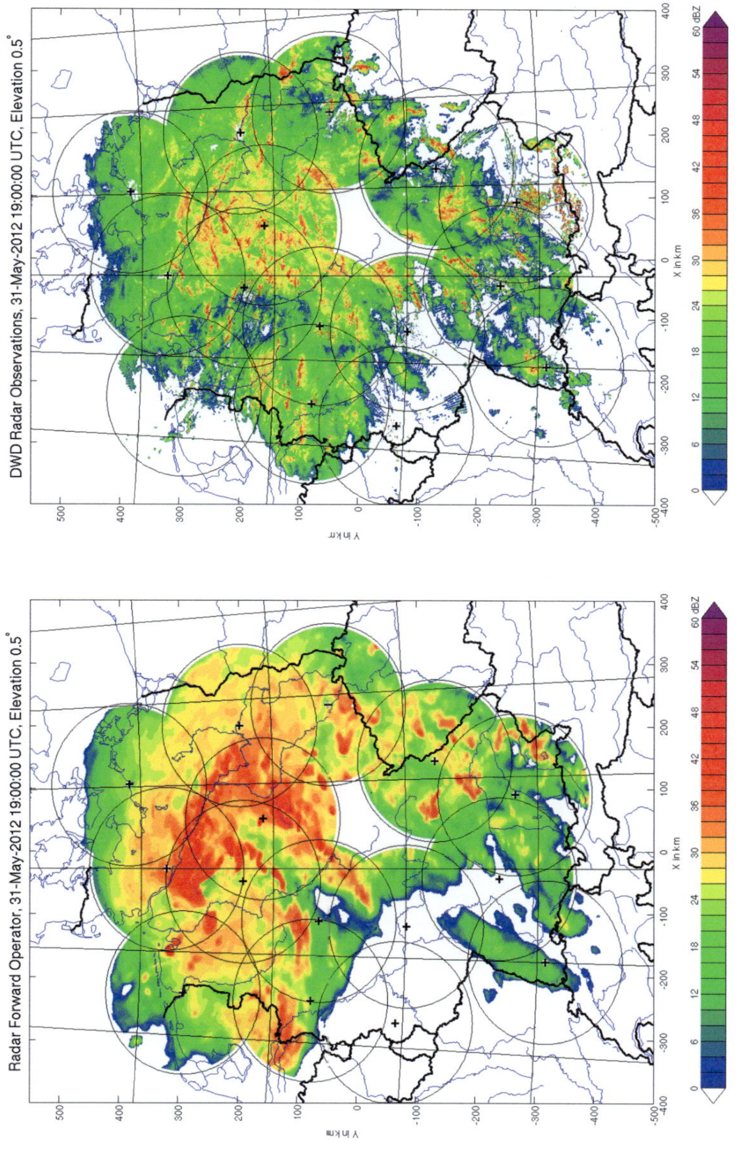

Figure 5.11: Same as Figure 5.10 but for 31.05.2012, 19 UTC.

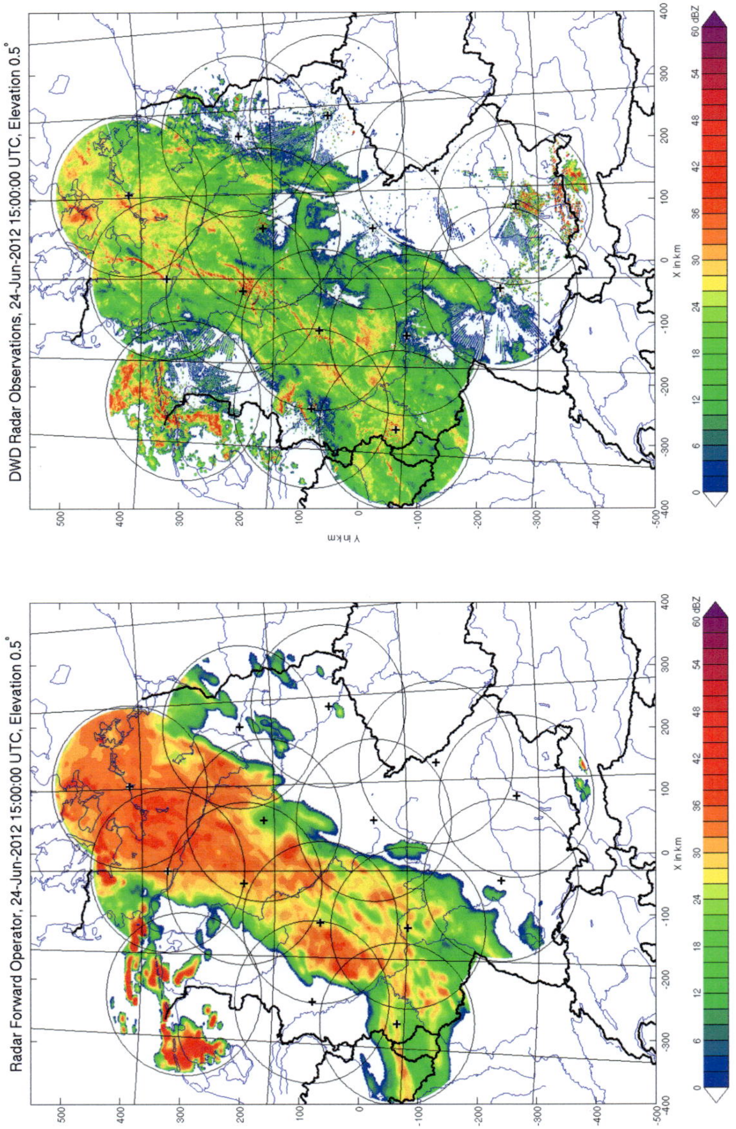

Figure 5.12: Same as Figure 5.10 but for 24.06.2012, 15 UTC.

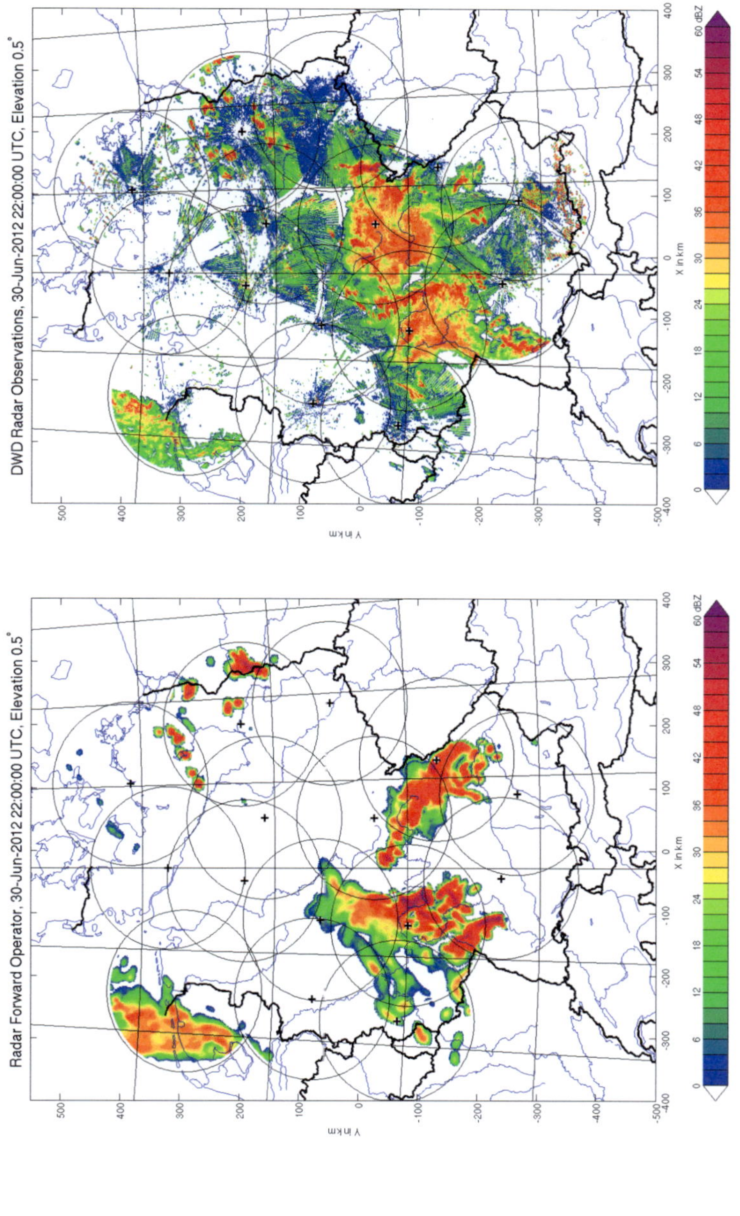

Figure 5.13: Same as Figure 5.10 but for 30.06.2012, 22 UTC.

5.7 Parallelisation and Vectorisation

In the end some short remarks on programming aspects of the radar forward operator should be made for the sake of completeness. Besides the requirement of good physical accuracy the operator also has to work with great computational efficiency when used operationally in the framework of data assimilation. The simulation program is embedded in the COSMO model and has to run on the vector-parallel supercomputer machines of the DWD, which currently are two NEC SX-9 clusters. Each cluster has 14 nodes with 16 processors and 8 replicated vector pipes. So the program code written in Fortan 90 and 95, respectively, (as well as the COSMO model) is fully vectorised.

Since the program runs parallel on several processors parallelisation strategies have to be applied, because each computational processor calculates a part of the domain only. The communication between the processors is provided by the message passing interface (MPI) which is linked to the programming code. The conventional method of the COSMO model is to divide the domain horizontally in commensurate rectangular subdomains and spread those equally on each processor. This is adopted by the radar forward operator and can be used when calculating the propagation of the radar beam using the equivalent earth model. However, this strategy can not be applied when simulating the propagation of the radar beam based on the actual simulated refractive index of air, since every calculation point along the radar beam has to know its preceding point of origin. So, one single radar beam cannot be subdivided on several processors. The solution is to cut the volume scan horizontally in "azimuthal slices" (resembling "pieces of cake") and spread them evenly over all processors. The method has been developed by Zeng (2013), where details can be found.

6 Cloud Microphysics Verification

6.1 Contoured Frequency by Altitude Diagrams (CFADs)

In the previous section the capabilities of the radar forward operator have been described and analysed, first by using idealised test cases and then by real case studies. One application of this operator is to compare the results of such real case simulations to corresponding radar observations, with the objective of verifying the COSMO model and its microphysical representation of precipitating clouds. The following discussion presents a method which can be proved beneficial for this purpose.

Note that in this thesis it is assumed that the observed radar data reflect the true meteorological event even though it is known that there are limitations and uncertainties of this measuring technique as described in chapter 4. A main advantage of using the radar forward operator is that some of these errors/characteristics can be simulated, and thus their individual effect on the radar observables can be examined. Furthermore, a better comparability to the observations is obtained. However, this topic is not considered in the following, since this last chapter only gives a first insight in the verification of the cloud microphysics.

Why CFADs?

The results of simulations of real synoptical situations compared to observations of the same atmospheric state measured by the radar network in Germany (introduced in section 4.4) are presented in Figures 5.10 to 5.13 in the previous section. The pictures of the radar composite help to get a first general idea of the differences and to make qualitative statements on the results of the numerical calculations. However, it is unlikely that the model predicts the precipitation at the exact same location (grid point) and the exact same time as a radar will measure it. Since the main interest in model verification is to find sytematical differences, it is inappropriate to contrast simulated data with observated data pointwise at each single coordinate point. Therefore, in order to facilitate a quantitative comparability a statistical methology is needed to display the data. One technique that fullfills this requirement are contoured frequency by altitude diagrams, subsequently abbreviated as CFAD(s). Yuter and Houze Jr. (1995) initially proposed this idea. This method is used here and will be described in detail in the following. Another positive aspect of CFADs is that they additionally provide information on the vertical structure of the radar data, which is useful since clouds and precipitation mainly form and develop in the vertical dimension.

What are CFADs?

A contoured frequency by altitude diagram is a diagram of the statistical horizontal distributions of a parameter that is vertically stratified (Yuter and Houze Jr., 1995). The graphic is a pseudo-three-dimensional view with the height given at the y-axis and the value of the parameter, in the following the radar reflectivity, given at the x-axis. The frequency of occurrence of the pairs of values are displayed as contours given in percent per dB per km. Figure 6.1 demonstrates the development and composition of such a diagram in three steps: (a) shows a simple historgam at a specific altitude (in the example 8 km); in (b) all histograms of all altitudes are combined in a perspective view and (c) is the actual CFAD, which is a simplified version of (b) converted in a topographic top projection with the described countour lines (here in intervals of $2.5\,\%\,\mathrm{dB}^{-1}\mathrm{km}^{-1}$).

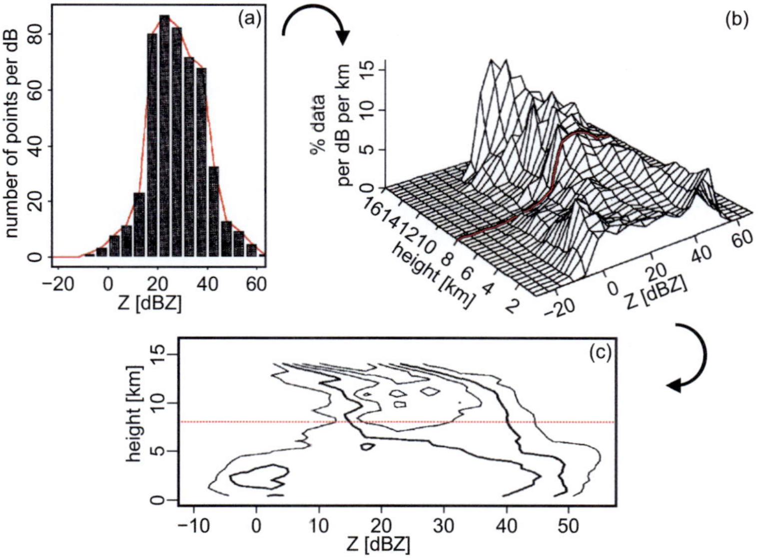

Figure 6.1: Developing process of a CFAD (Yuter and Houze Jr., 1995), starting with the histograms (a) at a specific altitude of 8 km, (b) for all heights and finally (c) as contour map, which is the actual CFAD in intervals of 2.5 $\%\,dB^{-1}km^{-1}$ contour lines. The thick lines are the 5 $\%\,dB^{-1}km^{-1}$ contour.

This method is not only able to avoid the temporal and spatial[13] mismatch between simulation and observation but also represents a picture of the volume of the radar data and thus allows to examine the characteristic frequency distribution of the radar reflectivity values as a function of height. In literature of cloud microphysics it is common use to distinguish between precipitation events from convective and stratiform storm structures. The characteristic distinctive features as seen by the radar are the intensity of precipitation and the size of the area covered by precipitation. Furthermore CFADs depict significant differences in the vertical structure of the radar parameters, which is drafted in Figure 6.2 for the reflectivity. In convective regions the predominant reflectivity values are higher than in stratiform regions. The convective profile remains constant or decreases with increasing altitude. In contrast, the stratiform profile first reveals a sharp maximum at a height of about 3 to 5 km, which corresponds to the temperature layer of 0 °C, then decreases as well as the convective profile. This significant maximum indicates the bright band (cf. section 4.3) that occurs in the melting layer of stratiform precipitation.

In the following all demonstrations are limited to the consideration of reflectivity. The same examinations can also be done with the radial velocity analogically (Yuter and Houze Jr., 1995).

How to create CFADs?

The output of the radar forward operator is not only restricted to the radar observables but also includes the simulated height above sea level. This configuration applies to all three methods of calculating the propagation of the radar beam. That simplifies the creation of CFADs since all required data are already given by the program. It is now important to distinguish the failed or missing reflectivity values between data with no information meaning that they would have been measured below the surface or shaded by the orography and data with

[13] when taking the collected data over a period of time to create the CFAD

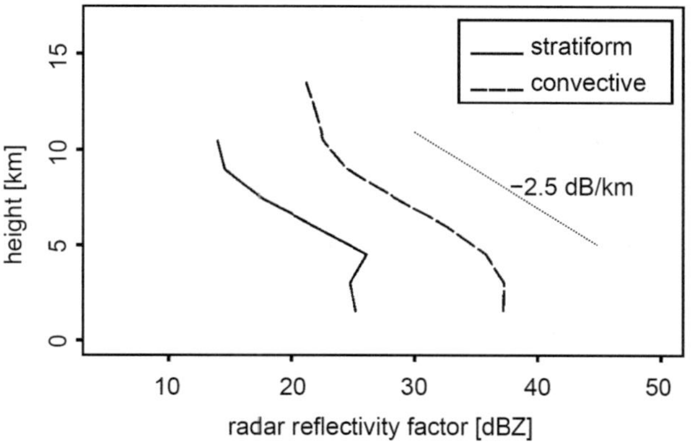

Figure 6.2: Characteristic mean vertical profiles of the radar reflectivity measured in a typical convective and stratiform precipitation event, respectively (Steiner et al., 1995).

no (or too little) reflectivity values since no (or not enough) hydrometeors exist in the corresponding volume of the radar pulse. The latter is a valuable information. So, all data with no information are set to "*not a number*" in order not to distort the dataset.

In the next step the actual values are restricted to a selected range. The relevant interval of the reflectivity has been chosen from 0.0 dBZ up to the maximum value. Lower values usually are associated with no precipitation and might even fall below the detection limit of the radar at larger ranges, depending on the quality of the radar receiving system. The lower limit of height is chosen as 0.4 km and 0.5 km, respectively (depending on Δh), since underneath there is not enough data because of the increasing altitude of the radar beam with increasing radial distance. The upper limit is set to 8.0 km, with some exceptions where the cloud top reaches higher altitudes (see first example of section 6.2).

Now, the increment step of the histograms has to be determined. In the following the bin widths are chosen to be $\Delta Z = 2\,\mathrm{dB}$ as well as $\Delta h = 200\,\mathrm{m}$ for stratiform cases and $\Delta h = 500\,\mathrm{m}$ for convective cases, respectively. The remaining pairs of values are then collected in a two-dimensional histogram and normalised in both directions. Finally a contour plot is produced with the resulting contour lines given in $\%\,\mathrm{dB}^{-1}\mathrm{km}^{-1}$.

With the procedure described above the already given data are processed unmodified to create CFADs for a start. However, even though the radar parameters and its corresponding height are directly provided by the radar forward operator another problem arises at once. These data are now given on a spherical grid in the coordinate system of the radar beam in terms of range, elevation and azimuth. Unfortunately the single grid points are irregularly distributed across the entire volume of coverage, meaning that there are many data points close to the radar station and on the contrary much less data far away from that. Figure 6.3 depicts all those measuring points that are collected by the radar during a complete volume scan. The plot demonstrates the nearer the single points are located to the radar station the smaller is the grid spacing. It also means that every grid point equates to a different effective resolution volume. Thus the resulting values of the radar reflectivity are dependent on the position of the radar station and the position of the precipitation at the time of the measurement, respectively, which is not desired. The aim of model verification is to find systematic differences in the model simulation compared to the measurement indepent of the coordinate system used or the relative position of the radar station to the precipitating cloud.

To demonstrate the consequences of a non-uniform grid spacing more clearly, an idealised test case as introduced in paragraph 5.6.1 is used (Figure 6.4). In this case, two radar stations are simulated contemporaneously. One station (II) is located directly nearby the cell. The position of the other station (I) is moved about 50 km in eastward direction. Figure 6.4 shows the respective ppi-plots of the lowest elevation measured by the two radar stations I and II, respectively, at the same time. The resulting reflectivity values look very similar at the surface as expected. With respect to the maximum range of a radar, of about 120 km, 50 km is no large distance. Anyway, as mentioned before, in the first attempt the

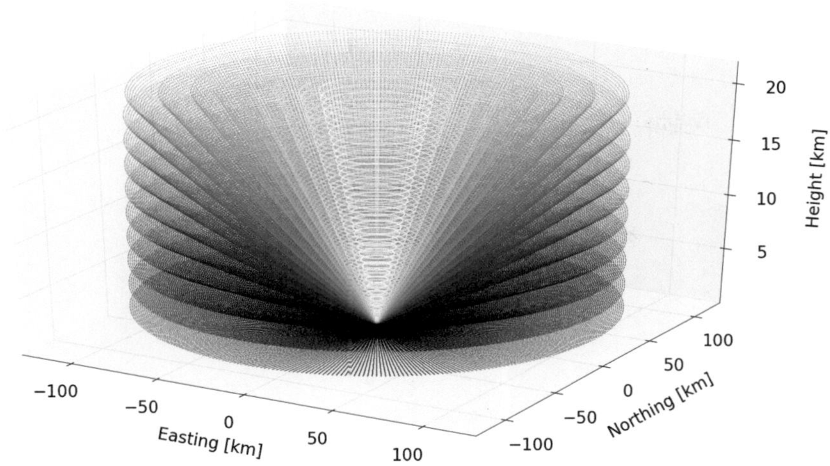

Figure 6.3: The complete set of measuring grid points provided by an entire volume scan of a radar. All radar observables are given on a spherical grid according to range, elevation and azimuth. The greyscales indicate the different elevation angles.

output of the radar forward operator is left unmodified to create the respective CFADs of the datasets for each radar station according to the above described procedure. This leads to significant different contour plots for the two radar stations as can be seen in Figure 6.5, because the data points far away from the station and in higher altitudes are underrepresented. In this case it is impossible to interpret the resulting vertical structure appropriately.

To avoid this problem there are two possibilities. The first method is applying at each pair of values a weight by the corresponding volume which it is representing. This is a simple and fast solution. But it solves the problem not entirely satisfying, as it addresses only the different coverage per measuring point, though the irregular grid distribution still remains. Figure 6.6 shows a strong improvement compared to Figure 6.5 but still reveals some minor differences between the both CFADs for I and II.

Transforming this data to a rectangular Cartesian grid would solve both problems, but requires interpolation. The most general way to linearly interpolate

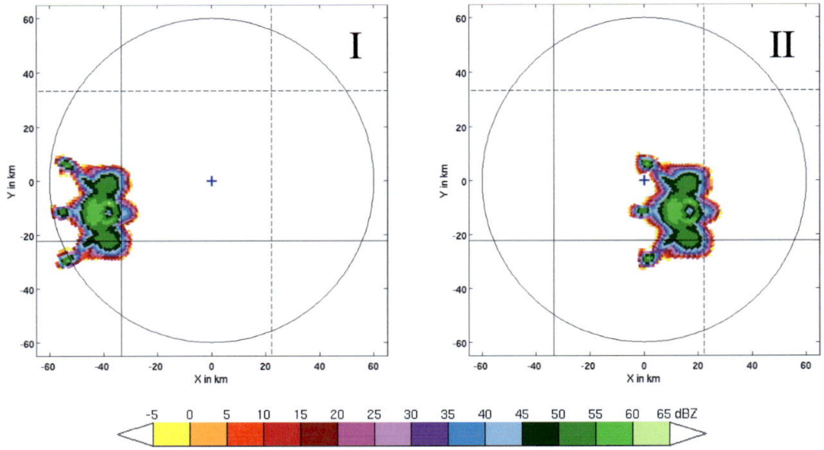

Figure 6.4: Two ppis of the lowest elevation from the view of two different radar stations (centre of the plots) I and II located at different positions. Each station measures the same atmospheric state at the same time using the ideal test case as described in paragraph 5.6.1. The plot is limited to half of the maximum range.

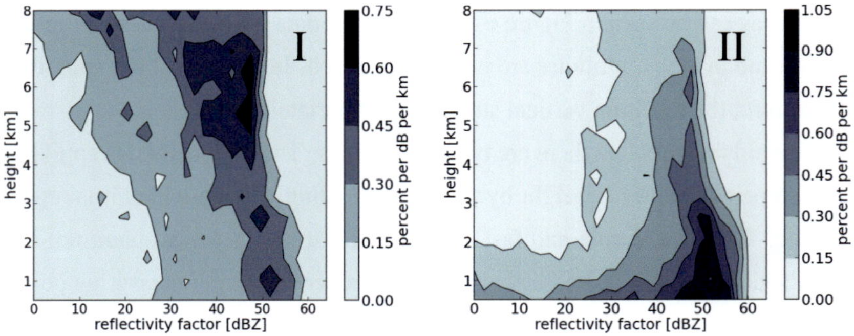

Figure 6.5: Resulting CFADs of the unmodified radar data on the polargrid for the radar stations I and II, respectively. The results are not comparable although the initial data describe the same meteorological situation at the same time. (For details see text.)

irregular gridded data is based on the Delaunay triangulation of the input data. Although there are already existing programming tools in different programming languages that provide this method (e.g. the method LinearNDInterpolator in the griddata program of the interpolation package in SciPy[14]) the calculation takes too much time to handle an entire set of reflectivity data of one volume scan. So this issue will not be addressed any further.

Instead of a triangular interpolation an alternative version is used here which is ascribed to Cressman (1959) and has been taken up and refined several times in the literature, e.g. by Langston et al. (2007). Another similar concept is the so-called k-nearest neighbour (knn) interpolation (e.g. Gao, 2009). In the following a combination of both methods is described.

First of all the spherical grid points of the radar observables defined on a polar coordinate system with respect to radial distance r, azimuth ϕ and elevation θ have to be expressed in terms of the Cartesian coordinates x_p, y_p and z_p, where z_p is already given by the height above mean sea level h,

$$\begin{aligned} x_p(r,\phi,\theta) &= s(r,\theta)\sin\phi \\ y_p(r,\phi,\theta) &= s(r,\theta)\cos\phi \\ z_p &= h. \end{aligned}$$

The arc length $s(r,\theta)$ is approximately described by the equation

$$s(r,\theta) = R_E \arcsin\left(\frac{r\cos\theta}{R_E+h}\right),$$

with the radius of the earth R_E. This projection is called an azimuthal equidistant projection and is locally nearly Cartesian.

[14]SciPy is short for scientific python.

In the next step "new" coordinate points along a Cartesian grid (x_c, y_c, z_c) are defined. The aim is to interpolate the simulated as well as the observed values of the radar reflectivity on each of these Cartesian grid points

$$Z_r(\vec{r}_p) = Z_r \begin{pmatrix} x_p \\ y_p \\ z_p \end{pmatrix} \quad \rightarrow \quad Z_{int}(\vec{r}_c) = Z_{int} \begin{pmatrix} x_c \\ y_c \\ z_c \end{pmatrix}.$$

The knn interpolation method as well as the Cressman method proposes for each Cartesian grid point $\vec{r}_{c,i}$ to pick all values of Z_r that are located in a specific circumcircle around that point. The radius of the circle d_{max} is set to 2.15 km[15], which means this is the maximum possible distance of a value that is to be regarded for the interpolation. When no data point lies within the described volume, the nearest located point is taken and d_{max} is set to the value of the distance of the second nearest located point. A weighting function w_i per grid point according to Cressman (1959) is provided by

$$w_i = \frac{d_{max}^2 - d_r^2}{d_{max}^2 + d_r^2},$$

where d_r is the distance between the polar grid point and the Cartesian grid point. The weighting function is unity if $d_r = 0$ and decreases to zero for $d_r \rightarrow d_{max}$. If $d_r > d_{max}$, it is set to zero. Finally the interpolation can be done according to following equation (Zhang et al., 2005)

$$Z_{int} = \frac{\sum w_i Z_{r,i}}{\sum w_i}.$$

The Figures 6.6 and 6.7 contain the corresponding CFADs according to the volume weighting method and the interpolation method, respectively. Both clearly show an improvement in contrast to Figure 6.5 where no modification of the data was done. These three pictures demonstrate that the interpolated values in Figure 6.7 reveal the best result with two nearly congruent contour plots. Therefore it is recommended to only use the interpolated data to create CFADs for further discussions.

[15] At this value for most Cartesian grid points there is at least one point of the spherical grid within its radius.

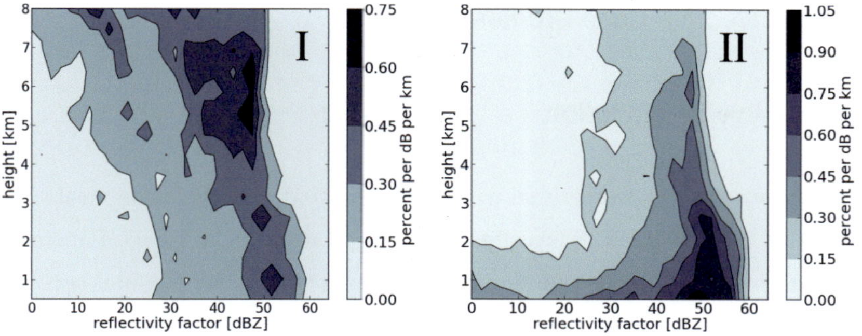

Figure 6.5: Resulting CFADs of the radar data on the polargrid and without volume averaging (revisited from page 80).

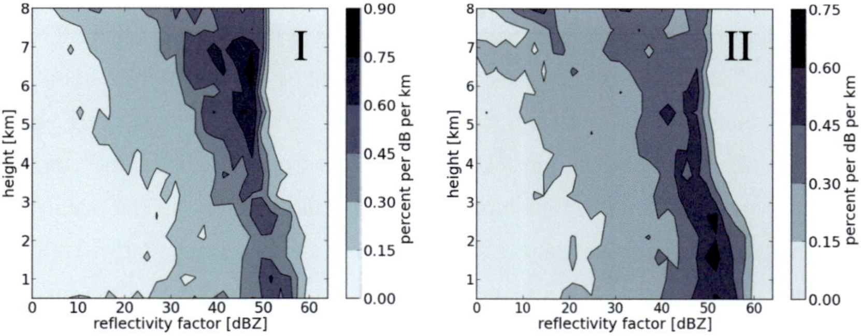

Figure 6.6: Same as Figure 6.5 but with volume averaging.

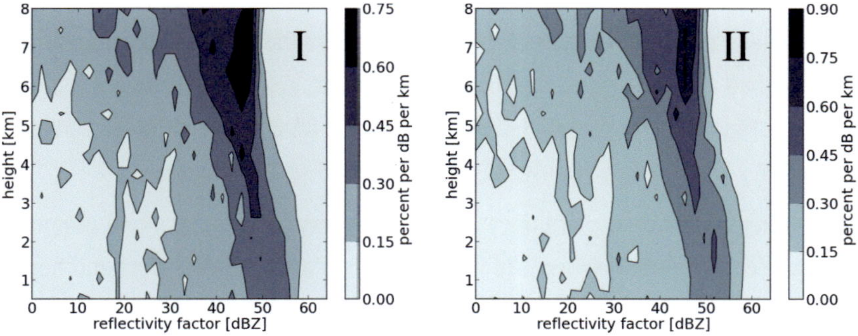

Figure 6.7: Same as Figure 6.5 but with the radar data interpolated on a cartesian grid.

6.2 Convective Case Studies

Convective Precipitation

The following two sections deal with some case studies of real synoptical situations as a first attempt to investigate the output of the radar forward operator in comparison with radar measurements. As indicated in the previous section, standard practise is to distinguish between convective and stratiform precipitation from a microphysical perspective, regarding intensity and amount of the atmospheric fallout (not to be confused with the classification of the hydrometeors themselves). This differentiation becomes also noticeable in the values of the radar reflectivity as can be seen in the vertical profiles (cf. Figure 6.2 on page 77). Convective events cause higher backscattering signals throughout all altitudes compared to stratiform events, which is attributed to the strong vertical velocities of the air within a convective cell. For this reason, the vertical motion w is the main distinguishing criterion for the separation. The occurring upwinds during such an event are at least $1\,\mathrm{m\,s^{-1}}$ and can reach up to $10\,\mathrm{m\,s^{-1}}$. Thereby large ice particles can be produced, since the motion of the air exceeds the terminal fall velocities v_t of most hydrometeors and they grow not only when falling but additionally during their updraft in higher (and colder) altitudes. This continues as long as the condition $|w| > v_t$ is given and can result in an up and down movement of the particles while they undergo an alternating melting and freezing process. These high vertical velocities also cause strong mixing of hydrometeors such that all types can be found in all altitudes and high reflectivity values can be constantly detected in several kilometres.

Another characteristic of convective precipitation is its spatial and temporal appearance in the atmosphere. In most cases it is a very local effect and hence strongly limited in its horizontal expansion while on the other hand the vertical extent is usually higher than in stratiform clouds. Also the lifetime of a convective event is strongly limited and does not last for more than a few hours. They often occur in the summertime. Houze Jr (1993) gives a more detailed physical description of convective precipitation.

First Example: 30th June 2012, 22 UTC

In paragraph 5.6.2 the results of some simulations of real meteorological situations were presented. A typical convective precipitation event happened on 30th June 2012 in Southern Germany. Figure 5.13 in the previous chapter on page 71 showed the lowest altitude of the measured and simulated values of the reflectivity at 22 UTC of that day. Two convective cells have formed simultaneously four hours earlier at the south-western border to the Alps and at the western border to France, moved towards the northwestern part of Germany and finally combined at 23 UTC resulting in heavy thunderstorms with strong rainfall over Hesse[16] and other parts of Germany. During the course of the storm extremely high reflectivity values occurred, extending over a particularly large area. This makes the event perfectly suitable for a first statistical comparison using CFADs.

For the creation of the contour plots the data of one radar station was used. In this case the radar station located in Offenthal (Hesse), near Frankfurt am Main, was chosen at the same point in time also shown in Figure 5.13. To demonstrate the selected location, Figure 6.8 depicts ppis of three elevations (0.5 °, 3.5°and 6.5°) of the simulation as well as the corresponding observation of this radar station at the time of examination. This choice was made because both the simulation and the observation detected the strongest values at the same time within the range of the same radar station and so there is a high spatial and temporal correlation between the datasets. This is an important fact when using the reflectivity values of only one radar at one specific point in time and still want to get a good and comparable statistic of both.

[16] `http://www.sueddeutsche.de/panorama/unwetter-in-nordhessen-blitz-erschlaegt-drei-frauen-auf-golfplatz-1.1397750`

Figures 6.9 and 6.10 compare the resulting CFADs of the simulated and the measured data, the first showing the complete vertical structure of the convective cell and the second only the lowest 8 km. It can be seen that even though the model overestimates the height of the cloud top compared to the observation the detected event still reveals an extreme large vertical extent. Moreover, the simulation shows a markedly broader distribution of the reflectivity values and especially in higher altitudes much more smaller values. However, on average the calculated values match the measurement. It is worth mentioning that (even though this is a clearly convective event) the structure of the CFADs (and ppis) also indicate the occurrence of a bright band.

Second Example: 31st May 2012, 15-16 UTC

A second example should confirm the results from the first example. So, another event that happened on 31st May 2012 starting at 15 UTC in Northern Germany was chosen where several single convective cells formed over the North Sea. In Figures 6.11 and 6.12, respectively, the ppi-plots of the reflectivity values are presented taken from the radar station in Emden (Lower Saxony) at 15 and 16 UTC. The station is located in the north-eastern part of Germany at the river mouth of the Ems into the North Sea. Here, also the melting layer is visible in the higher elevational ppi-plots of the measured data. The precipitation area is developed further to a large stratiform event that is examined in the second example in the next section.

The respective CFADs – a sum of the datasets of the two points in time – can be seen in Figure 6.13. Again, the simulated data show a broader distribution, mainly towards smaller reflectivity values. However, mostly higher values occur near ground. They also reach higher altitudes. So, both examples reveal similar differences between the simulation and measurements.

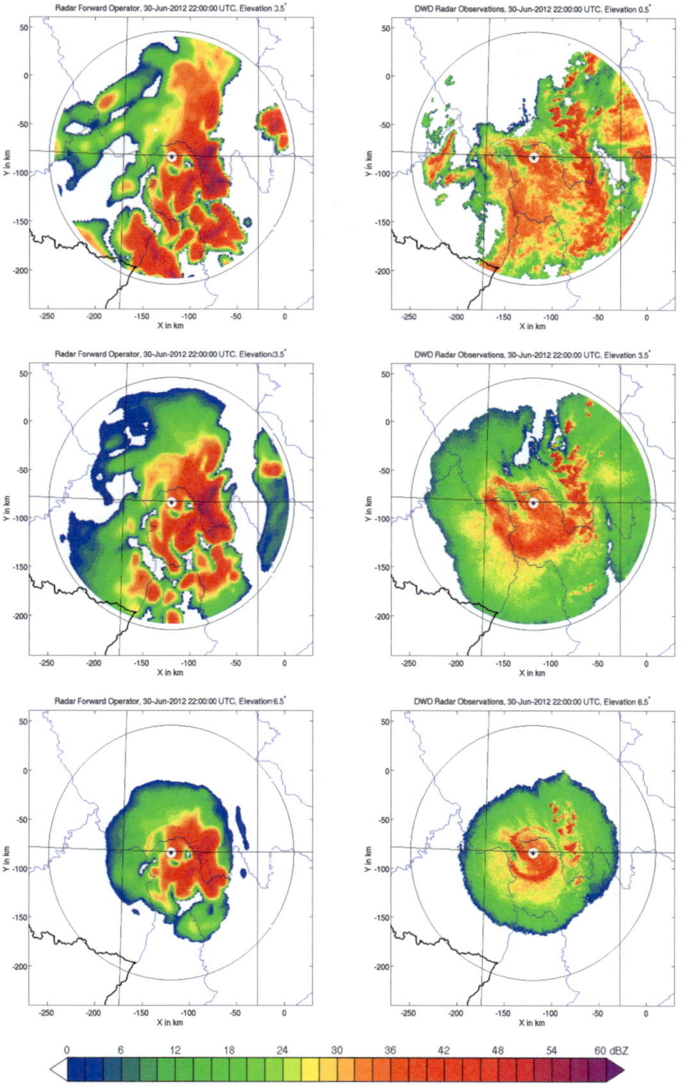

Figure 6.8: Simulated (left) and observed (right) radar reflectivity (in dBZ, see colour bar) for the first convective example: 30.06.2012, 22 UTC, radar station in Offenthal. Here, ppis of three different elevations (0.5°, 3.5°and 6.5°) are shown. For details see text.

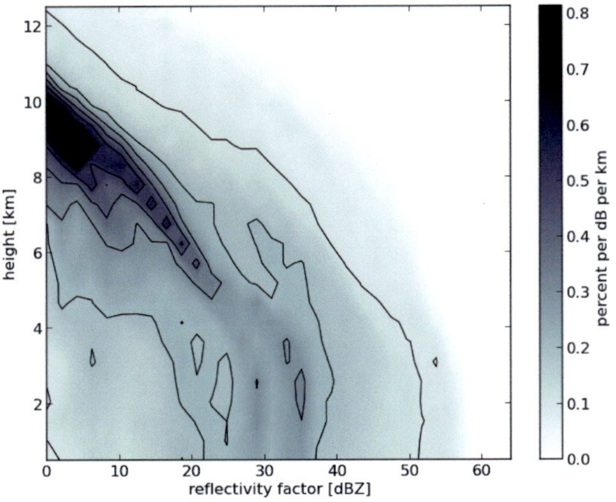

(a) Simulation.

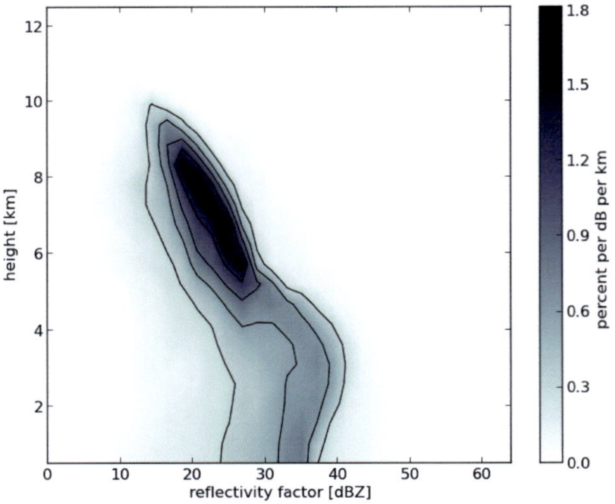

(b) Observation.

Figure 6.9: Resulting CFADs created from the data of the first convective example presented in Figure 6.8 (upper palet: from simulated data, lower plate: from observational data), showing the complete vertical structure. For details see text.

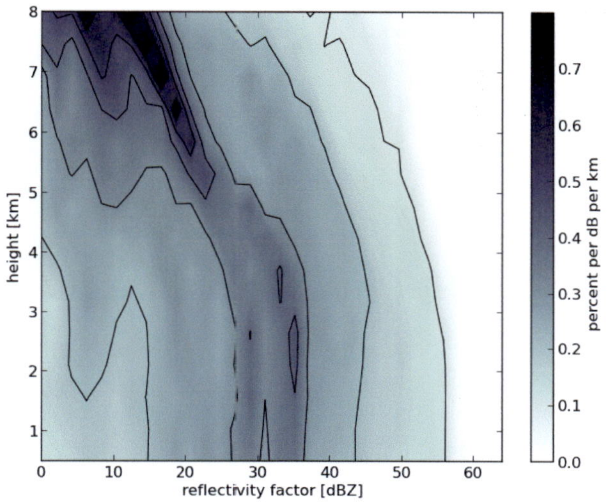

(a) Simulation.

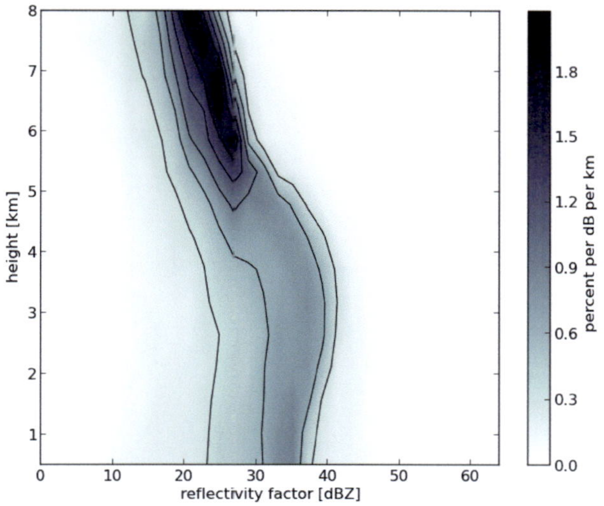

(b) Observation.

Figure 6.10: Same as Figure 6.9 but showing only the lower 8 km.

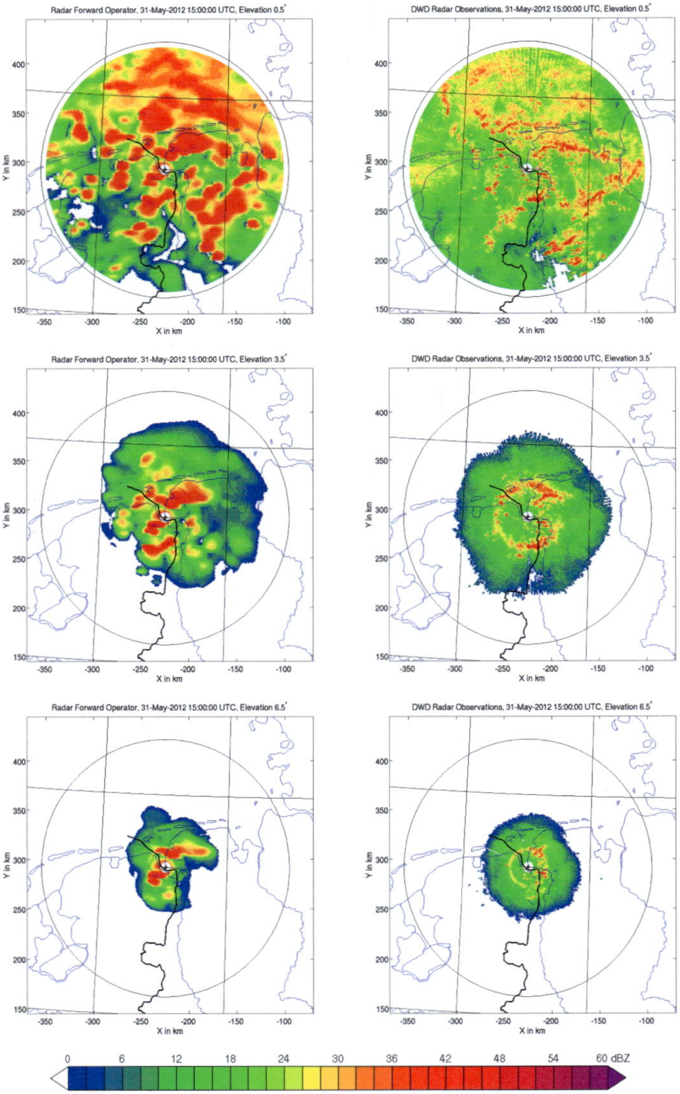

Figure 6.11: Same as Figure 6.8 but for the second convective example: 31.05.2012, 15 UTC, radar station in Emden. For details see text.

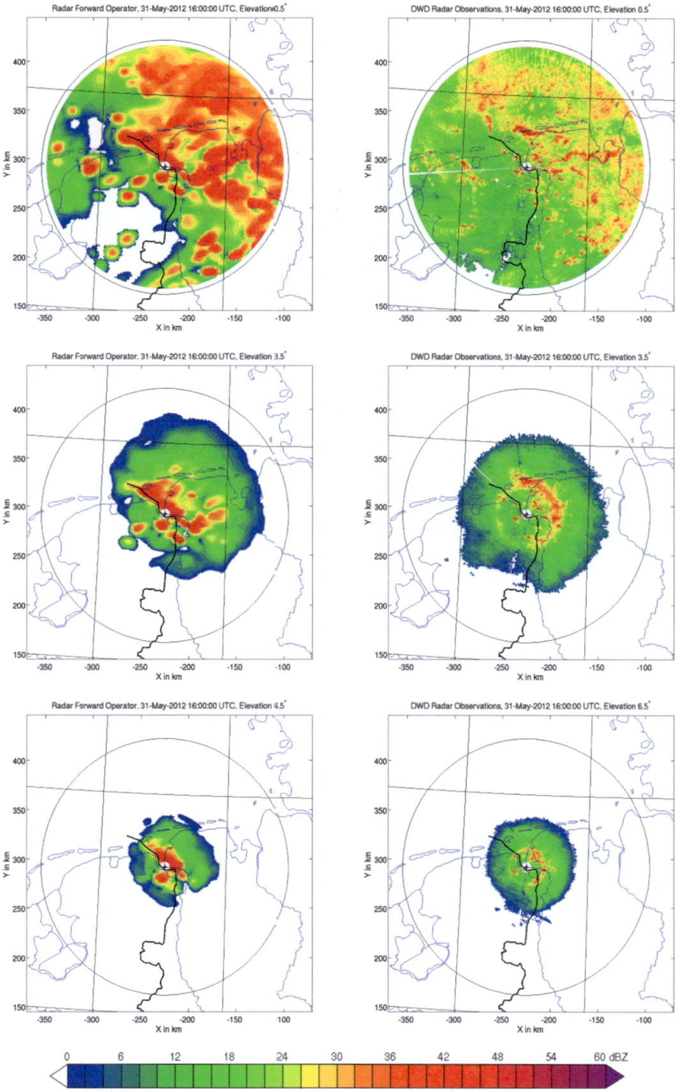

Figure 6.12: Same as Figur 6.11 but 1 h later: 31.05.2012, 16 UTC.

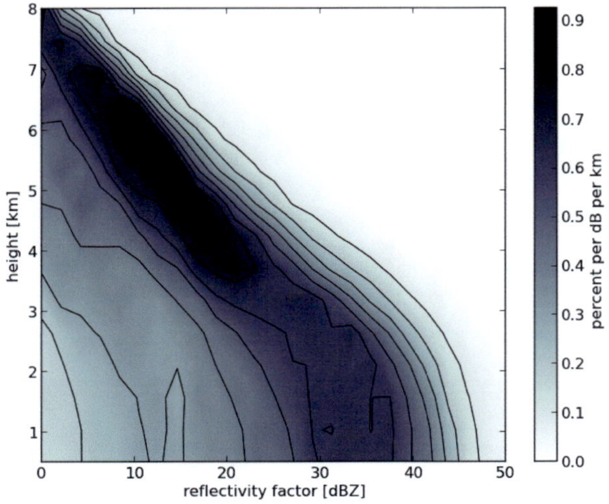

(a) Simulation.

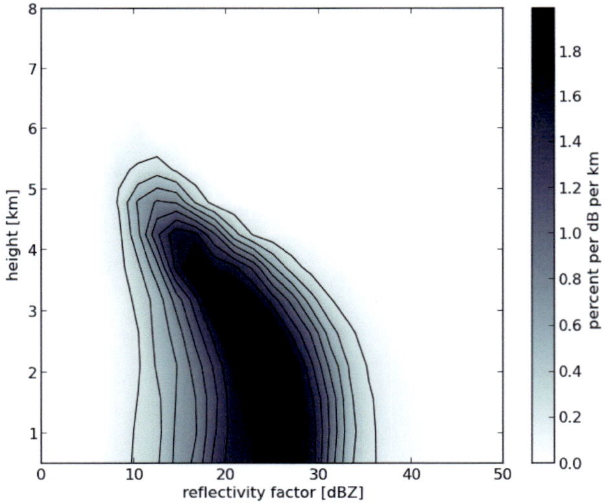

(b) Observation.

Figure 6.13: Same as Figure 6.9 but for the second convective example presented in Figure 6.11 and 6.12, respectively, showing the lower 8 km. The resulting CFADs are created from the sum of the data of both times. For details see text.

Third Example: 5th January 2012, 9-11 UTC

Another winterly convective event took place on 5th January 2012, when a strong squall line went through the detection area of a radar station, in this case the station located in Offenthal (Hesse) again. However, even though the occurring reflectivity values are lower than in the first example this particular event stands out by being very stable over a long period of time and was chosen as third example since precipitation with extremely high reflectivity values are quite rare. Furthermore it is remarkable that it happened in winter. The Figure 6.14 shows the ppis of the hourly output of the simulation and the corresponding observations from 9 to 11 UTC.

In this case three datasets are available to create the CFAD with an increased statistics. Figure 6.15 shows the resulting contour plot containing the data of every hour combined in one single diagram for both the calculated and the observed reflectivity values. Again the values of the simulation reach higher altitudes and in this case the measured data reveal a broader distribution especially near the ground but also a good agreement on average.

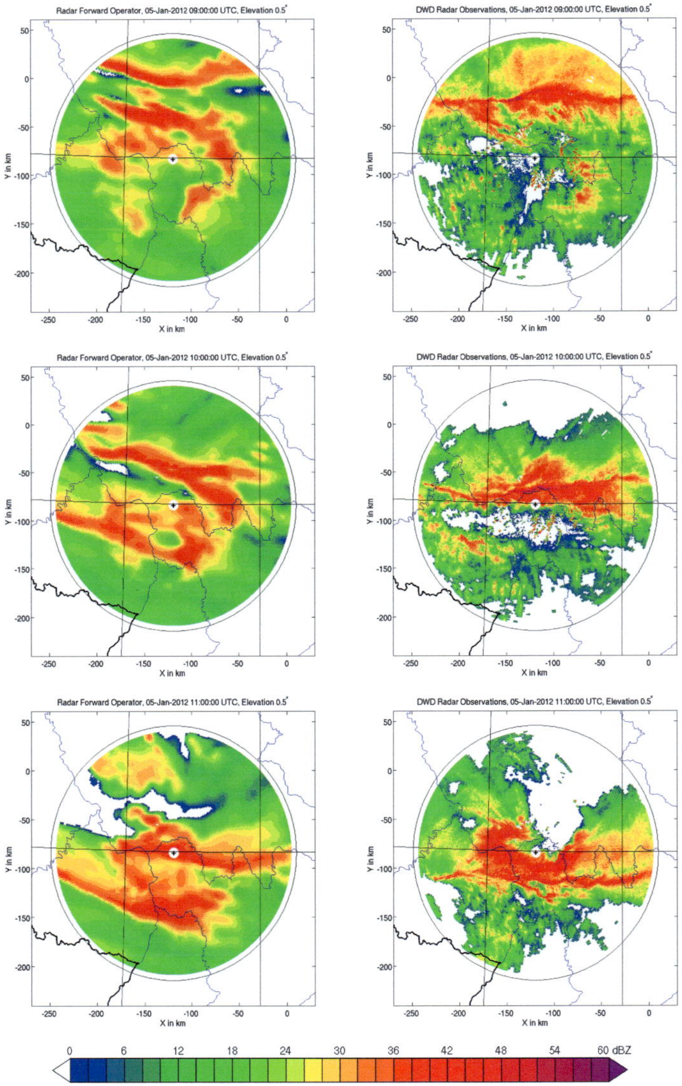

Figure 6.14: Same as Figure 6.8 but for the third convective example: 05.01.2012, 19 UTC, 10 UTC and 11 UTC, respectively, radar station in Offenthal. Here, three different times are plotted but with the lowest ppi (0.5°) only. For details see text.

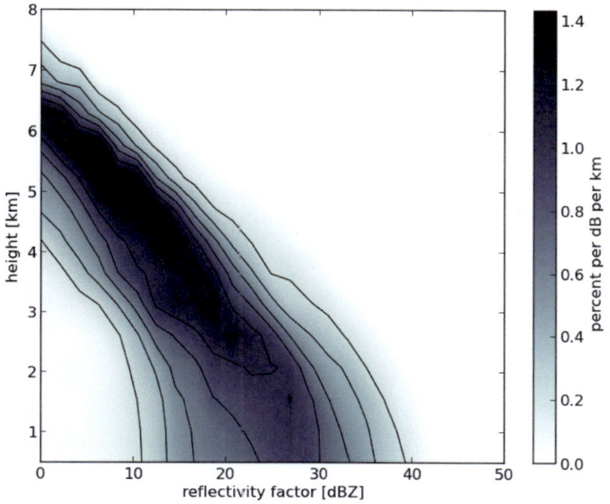

(a) Simulation.

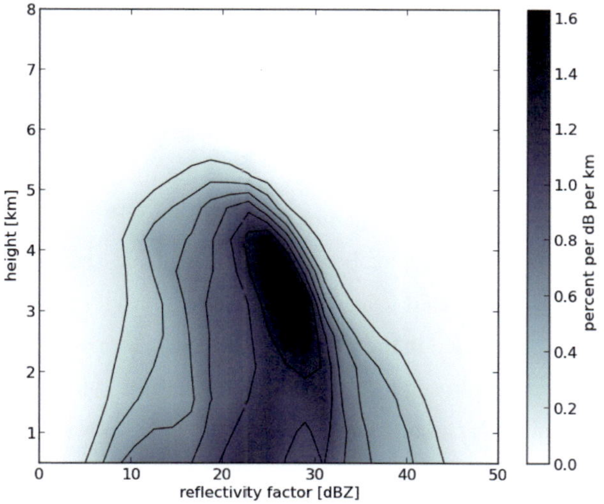

(b) Observation.

Figure 6.15: Same as Figure 6.13 but for the third convective example presented in Figure 6.14. Here, the sum of the data of three times is taken. For details see text.

6.3 Stratiform Case Studies

Stratiform Precipitation

Stratiform precipitation differs strongly from convective precipitation which is mainly due to the much lower vertical velocities occurring in stratiform clouds compared to convective cells. In a stratiform region the vertical motion of air does not exceed $1\,\mathrm{m\,s^{-1}}$, which equals the typical terminal fall speed of snow $v_{t,s}$ or ice crystals $v_{t,i}$. Generally it is claimed to be (Steiner et al., 1995; Houze Jr, 1993)

$$|w| \ll v_{t,\{s,i\}}.$$

This significantly affects the microphysical processes, since all hydrometeors are forced to grow exclusively during their falling through the cloud, there is no updraft of the particles. It also leads to a well separated distribution of the different types of hydrometeors, which means that snow and ice particles are found above the 0°C level and rain occurs predominantly below this layer, where the temperature exceeds the melting point. The transition region forms the so-called bright band (cf. section 4.3) which is characteristic of stratiform clouds, see Figure 6.2. It is also referred to as the melting layer, since most melting particles are found in this altitude.

Stratiform precipitation typically extends over a large area and can last up to several days. The detected values of the radar reflectivity are approximately homogeneously distributed in their horizontal extent. However, the rate of rainfall is much lower than in convective precipitation events. More information on stratiform precipitation can be found in Houze Jr (1993) as well.

First Example: 07th - 08th March 2012

The real case studies in the previous chapter (paragraph 5.6.2) also presented a typical stratiform precipitation event that happened on 07th March 2012 in Germany. The first rainfall was detected at 12 UTC in the north-western part of Germany over the North Sea and it took nearly 24 hours to move throughout the whole country from west to east as demonstrated in Figure 5.10 on page 68 showing the measured and simulated reflectivity values at 2 UTC on 08th March 2012. The precipitation extended nearly homogeneously over the entire region from north to south. This stable and consistent conditions are well suited to a stratiform case study.

Like in the previous section, only the data of one radar station were regarded for the creation of CFADs. At first, again the station in Emden (Lower Saxony) was chosen. The times of examination are 14 UTC as well as 16 UTC when the precipitation area hits the land surface, containing a lot of hydrometeors. Figures 6.16 and 6.17, respectively, show the corresponding ppis in the same manner as Figure 6.8 of the first example of a convective event. This selection was made for the same reasons described in the previous section. A preferably complete coverage of precipitation in the detection range of the radar, occurring in both the calculated and the measured data, is necessary to provide good and significant statistics using the data of just one radar station.

The corresponding CFADs are shown in Figure 6.18 as a sum of the two datasets of 14 UTC and 16 UTC. Since the reflectivity values do not reveal any significant differences between both points in time, the statistics could easily be increased. It is clearly visible that also in this case the simulated values reach higher altitudes than the observed values. Moreover, the average of the calculated radar reflectivity exceeds the mean reflectivity the radar measured by about 5 dB.

In a next step, a second radar station was selected that detected the event several hours later to create further CFADs of this event. At that time the stratiform precipitation field is still very extensive and stable but with extenuated intensity, see the lowest ppis of 2 UTC, 4 UTC and 6 UTC of the 08th March in Figure 6.19 (and Figure 5.10 for the complete radar composite of 2 UTC) in comparison to Figure 6.16 or 6.17 ten to 16 hours earlier. This second radar station is located in Dresden (Saxony) in Eastern Germany. The resulting CFADs

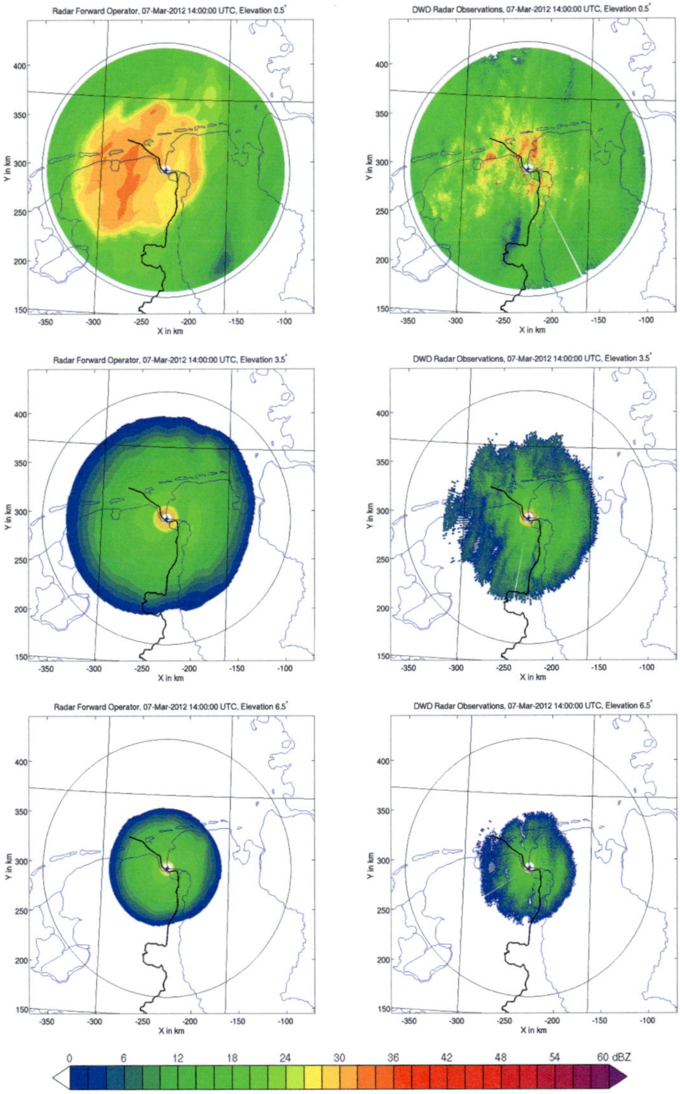

Figure 6.16: Simulated (left) and observed (right) radar reflectivity (in dBZ, see colour bar) for the first stratiform example: 07.03.2012, 14 UTC, for a first radar station in Emden. Here, ppis of three different elevations (0.5°, 3.5°and 6.5°) are shown. For details see text.

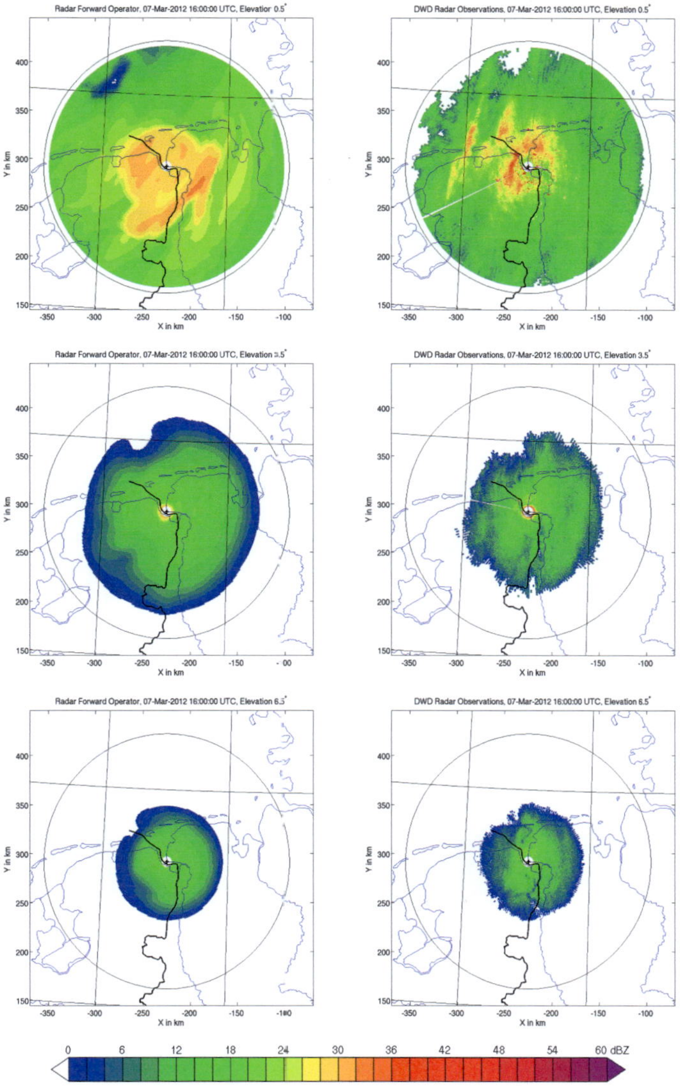

Figure 6.17: Same as Figure 6.16 but 2 h later: 07.03.2012, 16 UTC.

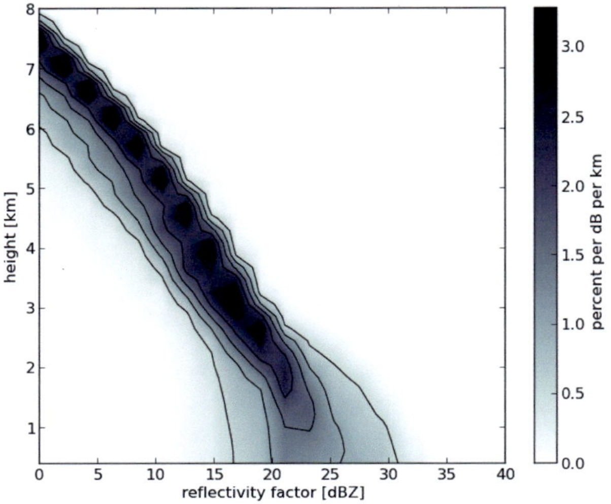

(a) Simulation.

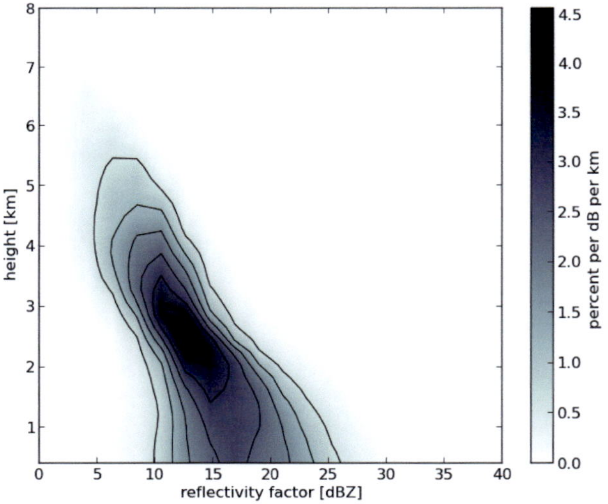

(b) Observation.

Figure 6.18: Resulting CFADs created from the data of the first stratiform example presented in Figure 6.16 and 6.17, respectively (upper palet: from simulated data, lower plate: from observational data), showing the lower 8 km. The CFADs are created from the sum of the data of three times. For details see text.

are shown in Figure 6.20, this time as a combination of the three sets of data. Since the precipitation is attenuated in its intensity, the vertical extent is also lower and hence it is sufficient to set the maximum height of the CFADs is to 5 km (instead of 8km). In this case calculated reflectivity values are much lower (less than 5 dB difference) than the measured values. However, the simulated precipitation extended to higher altitudes as found in all other cases before.

Second Example: 31st May 2012

Another interesting event happened on 31st May 2012 when a large stratiform precipitation area containing sporadic convective cores (cf. second example of a convective event in the previous section) covered nearly the whole country of Germany. The complete radar composite of 19 UTC is shown in Figure 5.11 on page 69. The 18 UTC and 19 UTC datasets of the radar station in Rostock (Mecklenburg-Vorpommern) in the north-eastern part of Germany were chosen for the creation of CFADs. Figures 6.21 and 6.22 show ppis of the three elevations. Of special interest is the significant occurrence of the bright band in the observational data which is clearly visible in the 3.5°and 6.5°elevation. This can be recognised by the concentric circle of higher reflectivity values around the centre of the radar station. The simulated data did not show this specific characteristic as clearly. However, the simulation registered already higher reflectivity values at ground level which could counteract the development of a bright band. In Figure 6.23 the combined CFADs of both times are depicted. The sharp maximum in the vertical structure which should indicate the bright band did not emerge as expected especially in the observational data. For this purpose more datasets collected over a long period of time (up to a month) would be necessary.

The ppis (Figure 6.24) and the corresponding CFADs (Figure 6.25) of the data taken from another radar station located in Prötzel (Brandenburg) in Eastern Germany at 18 UTC show the same results with increased significance and confirm what has already been stated.

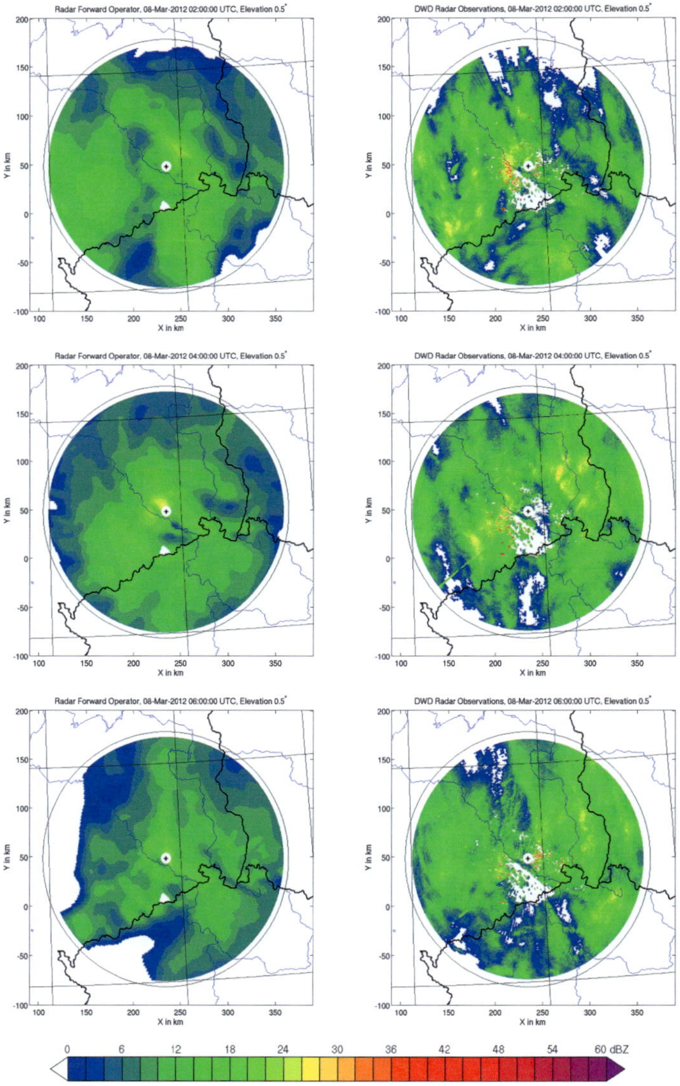

Figure 6.19: Same as Figure 6.16 but for the first stratiform example: 08.03.2012, 02 UTC, 04 UTC and 06 UTC, respectively, for a second radar station in Dresden. Here, three different times are plotted but with the lowest ppi (0.5°) only. For details see text.

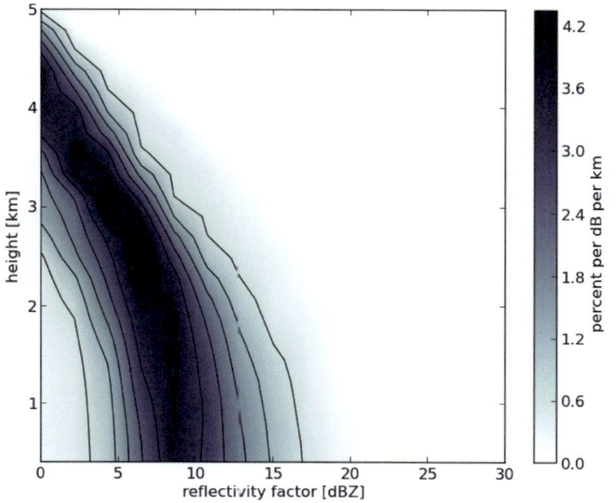

(a) Simulation.

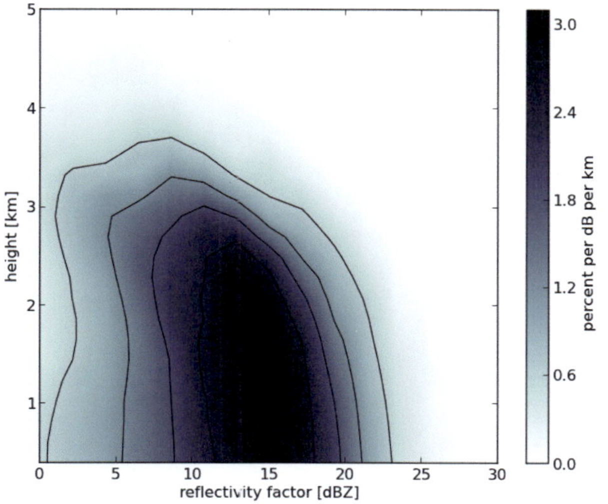

(b) Observation.

Figure 6.20: Same as Figure 6.18 but for the first stratiform example and second radar station presented in Figure 6.19, showing the lower 5 km. Here, the sum of the data of three times is taken. For details see text.

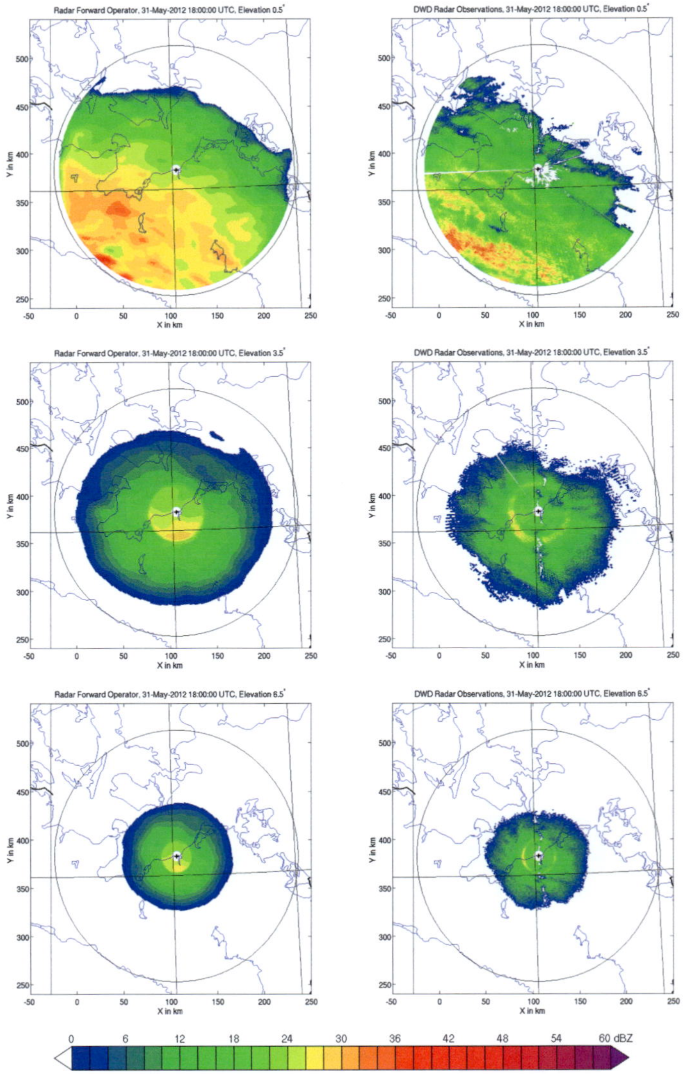

Figure 6.21: Same as Figure 6.16 but for the second stratiform example: 31.05.2012, 18 UTC, for a first radar station in Rostock. For details see text.

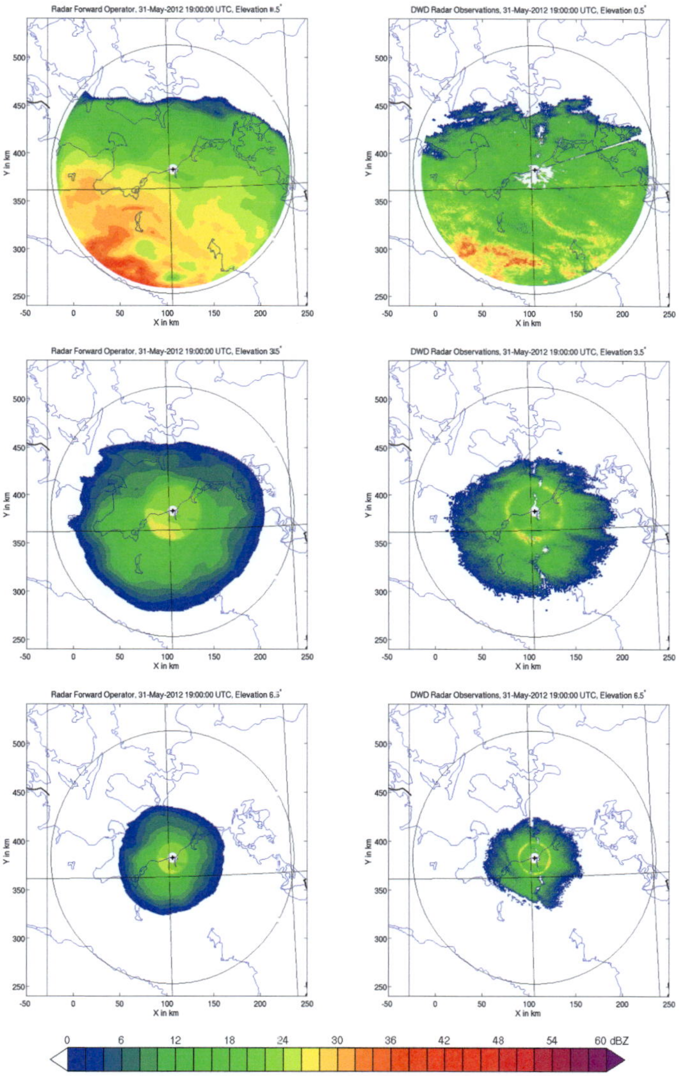

Figure 6.22: Same as Figure 6.21 but 1 h later: 31.05.2012, 19 UTC.

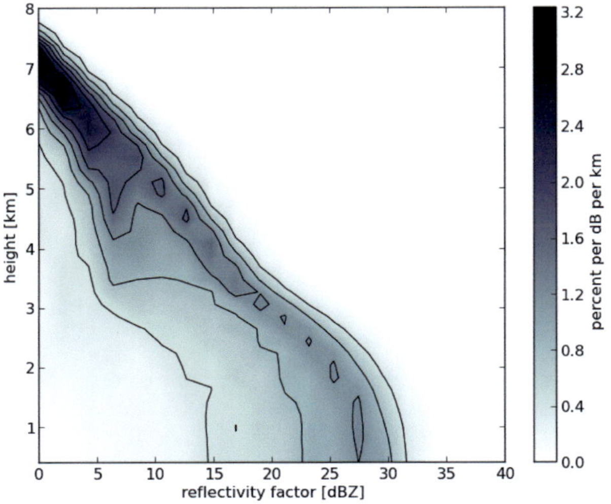

(a) Simulation.

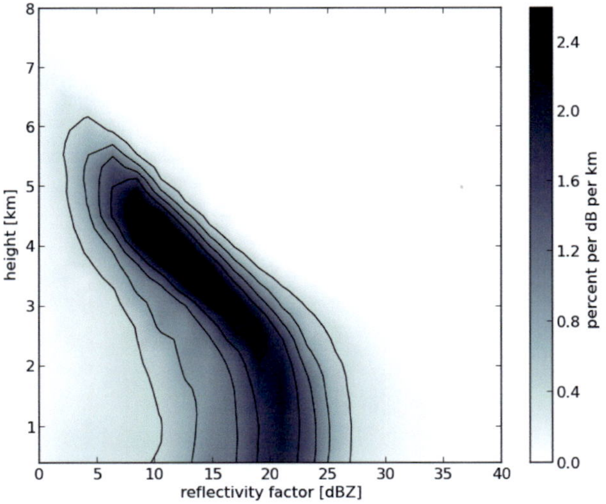

(b) Observation.

Figure 6.23: Same as Figure 6.18 but for the second stratiform example presented in Figure 6.21 and 6.22, respectively. For details see text.

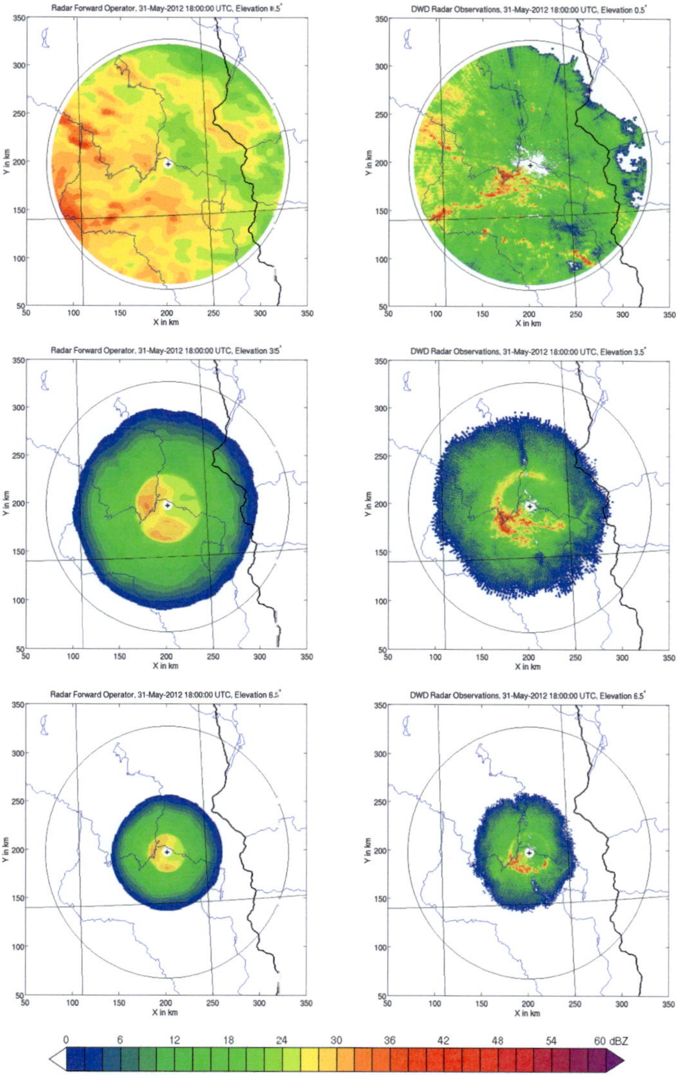

Figure 6.24: Same as Figure 6.21 but for the second stratiform example: 31.05.2012, 18 UTC, for a second radar station in Prötzel. For details see text.

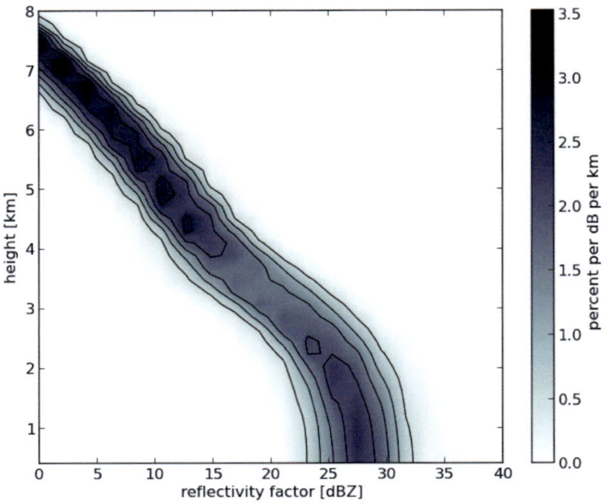

(a) Simulation.

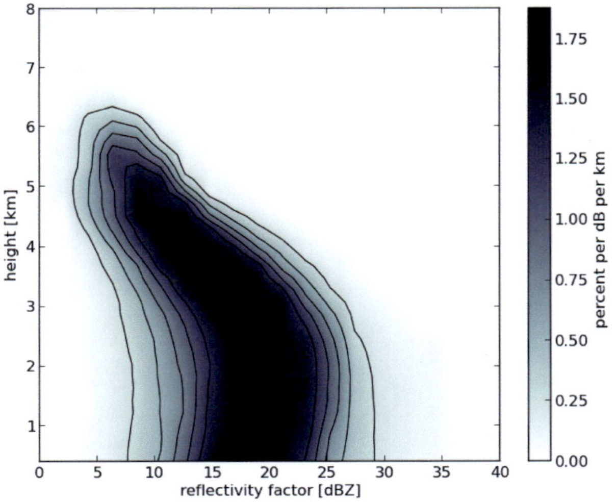

(b) Observation.

Figure 6.25: Same as Figure 6.23 but for the stratiform example and second radar station presented in Figure 6.24. Here, the data of only one time is taken. For details see text.

7 Conclusion and Outlook

The quantities measured by a weather radar system (reflectivity, radial velocity and polarisation parameters) do not directly represent those atmospheric variables that are predicted by numerical weather prediction (NWP) models (temperature, pressure, wind, total mass fraction of atmospheric water) and cannot directly be used to test the quality of the model forecast. However, having at hand variables predicted by a NWP model, which are appropriate to compute radar observables as reflectivity and radial velocity, synthetic radar data can be calculated. The numerical radar products thus obtained can be compared to 'real' radar measurements such that a direct measure of quality and reliability is available. This procedure comprises a so-called radar forward operator which is presented in this thesis. All important physical processes in the radar measurement are accounted for, each programmed in a separate module.

The development was done in close collaboration with Yuefei Zeng (2013) in a parallel PhD project. He concentrated on aspects of the electromagnetic wave propagation (such as the bending of the radar beam towards the earth's surface (Zeng et al., 2013) and the broadening of the radar pulse with distance) and on the vectorisation and parallelisation strategies which have to be applied in order to run the operator on supercomputer architectures. He further focused on the application of the operator for data assimilation.

The main part of this work consists of the simulation of the radar reflectivity using either the full Mie theory or the simpler Rayleigh approximation. In the former case the attenuation of reflectivity due to atmospheric hydrometeors can be taken into account additionally. Here, backscattering as well as extinction cross-sections are calculated separately for all (liquid, frozen and melting) types of hydrometeors and integrated over the particle size distributions as derived from model variables. The equivalent reflectivity and the extinction coefficient are initially simulated on the model grid and subsequently interpolated on spherical coordinates of the respective radar pulse. Afterwards, the attenuation can be calculated along the radar beam.

First idealised case studies show reasonable results and demonstrate the correct operation of the radar forward operator. Both possibilities of calculating the reflectivity, Mie theory (with or without attenuation) and Rayleigh, respectively, were tested. The reflectivity-attenuation-relation simulated from the model data compared to a theoretical consideration provides further evidence of a successful implementation. The influence of attenuation mostly becomes apparent towards smaller radar wavelengths, and hence should not be neglected. For this reason using the full Mie theory including extinction is preferable.

However, due to the complexity of the Mie theory, its calculation comes at higher computational cost compared to the Rayleigh approximation. To circumvent this problem, lookup tables are created for the reflectivity and the extinction coefficient when using the Mie theory. This has to be done individually for every hydrometeor class, i.e. for: rain, frozen snow, melting snow, frozen graupel and melting graupel, which altogether assemble 10 single tables. The radar effects of cloud water droplets and cloud ice crystals are still calculated using the Rayleigh approximation, since the Rayleigh condition (particle's diameter $\ll$ radar wavelength) is valid for small hydrometeors. The backscattering as well as the extinction cross-section are functions of the total mass fraction, the temperature and, in case of melting particles the degree of melting, and hence these quantities are the independent parameters of the lookup tables. The number of table nodes can be chosen arbitrarily. The tables are created during the first time step of the simulation (in a precalculation step), and in the following steps, the actual values for the equivalent reflectivity as well as the extinction coefficient are interpolated from the values contained in the lookup tables. With optimised configurations, this method performs even better than the simple approximation according to Rayleigh.

The complete radar forward operator is then applied to real meteorological situations. For comparison, measurements of up to 17 dual polarisation Doppler weather radar stations of DWD's radar network in Germany are used. Pseudo-composites in the form of ppi-plots of the reflectivity values at the lowest altitude of all radar stations are created. They allow a first qualitative evaluation of the precipitation forecast given by the model, which suggest that mostly the model is able the capture the meteorological situations quite well but also often overestimates the reflectivity values.

Investigations in model verification are the topic of the second part of this thesis, drawing further comparisons. In most cases the model does not predict precipitation events at the same time and the exact same position as the radar measures, and even slight differences in time and position prevent a meaningful comparison on the level of individual grid points. In order to reveal systematic discrepancies, a graphical representation of the radar data that enables a statistical comparison between the model and the observed data is prepared. This methodology is called contoured frequency by altitude diagram (CFAD), which was first proposed by Yuter and Houze Jr. (1995), and which is further investigated in this thesis. These contour plots focus on the statistics of the vertical structure of the radar data, which is an important information that describes main characteristics of clouds and precipitation developing mainly vertically by updrafts.

The output of the radar forward operator is given in spherical coordinates, because the radar measures its observables as a function of range, elevation and azimuth. It is examined whether these irregularly distributed grid points and accordingly the different grid spacing influence the statistical evaluation of the radar data by CFADs. Two radar stations detecting the same idealised precipitation area from different distances were simulated. The resulting CFADs reveal significant differences since the data near the radar station are overrepresented compared to the data further away from it. Secondly, every given value represents a different effective resolution volume, which can be adjusted by a volume weighting. However, the best way to avoid both problems is to interpolate the reflectivity values on a Cartesian grid, resulting in almost equal CFADs of the same event from radar stations at different locations. This is an important step to do, otherwise the interpretation of the contour plots can be misleading.

Convective and stratiform examples show the applicability of this methodology to comparative studies. This was done using the data of individual events detected by one radar station. However, more statistics (longer timescales, complete radar composite) would have to be used in order to make reliable statements on systematic differences between model output and the observations. Nevertheless, the CFADs of individual stations already show some significant discrepancies. In all cases the model overestimates the cloud top height compared to radar measurements. Also, the simulated reflectivity values are typically too high in regions of precipitation (rain) near the ground, while at higher altitudes much more values of low reflectivity occur in the simulation. The remaining challenge is to figure out where theses differences might originate from (e.g. from the choice of the particle size distribution of the different hydrometeor types within the COSMO model or from the general properties of the cloud microphysical scheme). But this is beyond the scope of this thesis. Furthermore, an interesting question would be how different configurations of the radar forward operator and the impacts of individual physical aspects in the radar simulator affect the CFADs, in order to find the best compromise between accuracy and computational costs in the simulation.

This radar forward operator has been developed for the reflectivity and the radial velocity up to now. However, there is also a set of polarisation parameters that can be detected with the radar technology now installed by DWD. The simulation of these parameters requires a lot more effort and is still an open project for the future. Additional information about the hydrometeors' shape and its angular orientation relative to the incident radiation (canting angle) is needed, that is not provided by the model and hence has to be somehow prescribed. A simulation of the polarisation parameters within the radar forward operator can be done, e.g. by building on parts of the program SynPolRad (Pfeifer, 2007; Pfeifer et al., 2008). Here, also expensive computations are involved which will require the extensive use of lookup tables.

Bibliography

Baldauf, M. et al. (2011). Operational Convective-Scale Numerical Weather Prediction with the COSMO Model: Description and Sensitivities. *Monthly Weather Review*, 139(12):3887–3905.

Bean, B. R. and Dutton, E. J. (1966). *Radio Meteorology*. National Bureau of Standards Monograph 92, Washington D.C.

Blahak, U. (2004). *Analyse des Extinktionseffektes bei Niederschlagsmessungen mit einem C-Band Radar anhand von Simulation und Messung*. PhD thesis, University of Karlsruhe.

Blahak, U. (2008). An Approximation to the Effective Beam Weighting Function for Scanning Meteorological Radars with an Axisymmetric Antenna Pattern. *Journal of Atmospheric and Oceanic Technology*, 25(7):1182–1196.

Blahak, U. (2009). RADAR_MIE_LM and RADAR_MIELIB – Calculation of Radar Reflectivity from model output. Technical report, IMK, Karlsruhe, Germany. available on demand.

Bohren, C. F. and Huffman, D. R. (1983). *Absorption and scattering of light by small particles*. John Wiley and Sons, New York.

Bringi, V. N. and Chandrasekar, V. (2001). *Polarimetric Doppler weather radar: principles and applications*. Cambridge University Press, Cambridge.

Bruggemann, D. A. G. (1935). Berechnung verschiedener physikalischer Konstanten von heterogenen Substanzen I. Dielektrizitätskonstanten und Leitfähigkeiten der Mischkörper aus isotropen Substanzen. *Annalen der Physik*, 416(7):636–664.

Caumont, O. et al. (2006). A radar simulator for high-resolution nonhydrostatic models. *Journal of Atmospheric and Oceanic Technology*, 23(8):1049–1067.

Cressman, G. P. (1959). An operational objective analysis system. *Monthly Weather Review*, 87(10):367–374.

Doms, G. (2011). A Description of the Nonhydrostatic Regional COSMO-Model, Part I: Dynamics and Numerics. Technical report, Deutscher Wetterdienst, Offenbach, Germany. Available online at `http://cosmo-model.org/`.

Doms, G. et al. (2011). A Description of the Nonhydrostatic Regional COSMO-Model, Part II: Physical Parameterization. Technical report, Deutscher Wetterdienst, Offenbach, Germany. Available online at `http://cosmo-model.org/`.

Doviak, R. J. and Zrnic, D. S. (1993). *Doppler Radar and Weather Observations*. Academic Press, Inc., San Diego.

Field, P. R., Heymsfield, A. J., and Bansemer, A. (2007). Snow Size Distribution Parameterization for Midlatitude and Tropical Ice Clouds. *Journal of the Atmospheric Sciences*, 64(12):4346–4365.

Gao, F. (2009). Fast k-Nearest-Neighbors Calculation for Interpolation of Radar Reflectivity Field. *Journal of Atmospheric and Oceanic Technology*, 26(7):1410–1414.

Gunn, K. L. S. and Marshall, J. S. (1952). Measurement of Snow Parameters by Radar. *Journal of Meteorology*, 9(5):322–327.

Handwerker, J. (2002). Cell tracking with TRACE3D – A new algorithm. *Atmospheric Research*, 61(1):15–34.

Hartree, D. R., Michel, J. G. L., and Nicolson, P. (1946). Practical methods for the solution of the equations of tropospheric refraction. *Meteorological factors in radio wave propagation*, pages 127–168.

Houze Jr, R. A. (1993). *Cloud dynamics*, volume 53 of *International Geophysics Series*. Academic Press, Inc., San Diego.

Järvinen, H. et al. (2009). Doppler radar radial winds in HIRLAM. Part I: observation modelling and validation. *Tellus A*, 61(2):278–287.

Kerker, M. (1969). *The scattering of light, and other electromagnetic radiation*. Academic Press, Inc., New York.

Kessler, E. (1969). *On Distribution and Continuity of Water Substance in Atmospheric Circulation*. Meteorological Monograph Series. American Meteorological Society, Boston.

Langston, C., Zhang, J., and Howard, K. (2007). Four-dimensional dynamic radar mosaic. *Journal of atmospheric and oceanic technology*, 24(5):776–790.

Liebe, H. J., Hufford, G. A., and Manabe, T. (1991). A model for the complex permittivity of water at frequencies below 1 THz. *International Journal of Infrared and Millimeter Waves*, 12(7):659–675.

Lindskog, M. et al. (2004). Doppler Radar Wind Data Assimilation with HIRLAM 3DVAR. *Monthly Weather Review*, 132(5):1081–1092.

Marshall, J. S. and Palmer, W. M. (1948). The Distribution of Raindrops with Size. *Journal of Meteorology*, 5(4):165–166.

Mätzler, C. (1998). Microwave properties of ice and snow. *Solar System Ices, Astrophysics and Space Science Library*, 227:241–257.

Maxwell-Garnett, J. C. (1904). Colours in metal glasses and in metallic films. *Proceedings of the Royal Society of London*, A203:385–420.

Mie, G. (1908). Beiträge zur Optik trüber Medien, speziell kolloidaler Metallösungen. *Annalen der Physik*, 330(3):377–445.

Noppel, H. et al. (2006). A two-moment cloud microphysics scheme with two process-separated modes of graupel. In *12th AMS Conference on Cloud Physics*, Madison, Wisconsin.

Oguchi, T. (1983). Electromagnetic Wave Propagation and Scattering in Rain and Other Hydrometeors. *Proceedings of the IEEE*, 71(9):1029–1078.

Pfeifer, M. (2007). *Evaluation of precipitation forecasts by polarimetric radar*. PhD thesis, Ludwig-Maximilians-Universität München.

Pfeifer, M. et al. (2008). A polarimetric radar forward operator for model evaluation. *Journal of Applied Meteorology and Climatology*, 47(12):3202–3220.

Puskeiler, M. (2013). *Radarbasierte Analyse der Hagelgefaährdung in Deutschland*. PhD thesis, Karlsruhe Institute of Technology.

Ray, P. S. (1972). Broadband complex refractive indices of ice and water. *Applied Optics*, 11(8):1836–1844.

Rogers, R. R. (1979). *A Short Course in Cloud Physics*. Pergamon Press, Oxford.

Ruffieux, D. and Illingworth, A. J. (2013). Final Report. Technical report, EG-CLIMET. Available online at `http://wiki.eg-climet.org/index.php?title=Final_Report`.

Rutledge, S. A. and Hobbs, P. V. (1984). The Mesoscale and Microscale Structure and Organization of Clouds and Precipitation in Midlatitude Cyclones. XII: A Diagnostic Modeling Study of Precipitation Development in Narrow

Cold-Frontal Rainbands. *Journal of the Atmospheric Sciences*, 41(20):2949–2972.

Sauvageot, H. (1992). *Radar Meteorology*. Artech House, Inc., Boston.

Schättler, U., Doms, G., and Schraff, C. (2012). A Description of the Non-hydrostatic Regional COSMO-Model, Part IIV: User's Guide. Technical report, Deutscher Wetterdienst, Offenbach, Germany. Available online at `http://cosmo-model.org/`.

Seifert, A. (2002). *Parametrisierung wolkenmikrophysikalischer Prozesse und Simulation konvektiver Mischwolken*. PhD thesis, University of Karlsruhe.

Seifert, A. and Beheng, K. D. (2006). A two-moment cloud microphysics parameterization for mixed-phase clouds. Part 1: Model description. *Meteorology and Atmospheric Physics*, 92(1-2):45–66.

Skolnik, M. I. (1980). *Introduction to Radar Systems*. McGraw-Hill, Inc., New York.

Steiner, M., Houze Jr., R. A., and Yuter, S. E. (1995). Climatological characterization of three-dimensional storm structure from operational radar and rain gauge data. *Journal of Applied Meteorology*, 34(9):1978–2007.

Steppeler, J. et al. (2003). Meso-gamma scale forecasts using the nonhydrostatic model LM. *Meteorology and Atmospheric Physics*, 82(1-4):75–96.

Turton, J. D., Bennetts, D. A., and Farmer, S. F. G. (1988). An introduction to radio ducting. *Meteorological Magazine*, 117(1393):245–254.

Ulbrich, C. W. (1983). Natural Variations in the Analytical Form of the Raindrop Size Distribution. *Journal of Climate and Applied Meteorology*, 22(10):1764–1775.

Warren, S. G. (1984). Optical constants of ice from the ultraviolet to the microwave. *Applied Optics*, 23(8):1029–1078.

Weisman, M. L. and Klemp, J. B. (1982). The Dependence of Numerically Simulated Convective Storms on Vertical Wind Shear and Buoyancy. *Monthly Weather Review*, 110(6):504–520.

Yuter, S. E. and Houze Jr., R. A. (1995). Three-dimensional kinematic and microphysical evolution of Florida cumulonimbus. Part II: Frequency distributions of vertical velocity, reflectivity, and differential reflectivity. *Monthly Weather Review*, 123(7):1941–1963.

Zeng, Y. (2013). *Efficient Radar Forward Operator for Operational Data Assimilation within the COSMO-model*. PhD thesis, Karlsruhe Institute of Technology.

Zeng, Y. et al. (2013). Radar Beam Tracing Methods Based on Atmospheric Refractive Index. submitted manuscript.

Zhang, J., Howard, K., and Gourley, J. J. (2005). Constructing three-dimensional multiple-radar reflectivity mosaics: Examples of convective storms and stratiform rain echoes. *Journal of Atmospheric and Oceanic Technology*, 22(1):30–42.

Danksagung

Der Letzte macht das Licht aus.

Bevor ich als letztes verbliebenes Mitglied der Arbeitsgruppe “Wolkenphysik und Radarmeteorologie” das letzte Kapitel beende, möchte ich noch einigen Personen danken, die mich in den vergangenen Jahren begleitet und zum Gelingen dieser Arbeit beigetragen haben.

Mein erster Dank geht an Prof. Klaus Dieter Beheng, der mir die Möglichkeit gab in seiner Arbeitsgruppe zu promovieren und außerdem für den nötigen und motivierenden Antrieb gegen Ende meiner Promotion sorgte. Vielen Dank für die fürsorgliche Betreuung auch über den Einstieg in den Ruhestand hinaus.

Ich bedanke mich bei Michael Kunz für die bereitwillige Übernahme des Korreferats.

Ganz großer Dank gilt meinem direkten Betreuer Ulrich Blahak für die unzähligen Autostunden zwischen Offenbach und Karlsruhe und die vielen Telefonate (ich weiß, es hätten manchmal mehr sein können) und E-Mails, mit denen er es schaffte, trotz der Entfernung, jederzeit – und mit schier unendlichem Wissen – mit Rat und Tat zur Seite zu stehen.

Meinem Bürokollegen und Kompagnon Yuefei Zeng danke ich für die gute Zusammenarbeit und die angenehme Gesellschaft, mit der er in seiner Art immer wieder das Vorurteil bestätigte, dass man Chinesen einfach mögen muss. So, I like chinese...

Besonderer Dank geht an meinen Ex-Büronachbarn Jan Handwerker, der mir bis zuletzt ein treuer Begleiter zur Kantine war und auch ab und zu mit mir den Arbeitsweg geteilt hat. Darüber hinaus, hatte ich das Vergnügen mit ihm u.a.

gemeinsame Radtouren zu bestreiten – die Elberadtour plane ich nach wie vor fest ein – und Spiele des KSC zu besuchen.

Gabi Klinck und Jan Handwerker sorgten stets für einen funktionierenden Arbeitsrechner und versorgten mich mit allerlei Sonderwünschen, wie Inkscape, Python und sogar Skype. Vielen Dank auch an Gabi, die sich zuverlässig auf meine Seite stellte, wenn Jan mal wieder damit drohte mir die Quota zu kürzen.

Der gesamten Arbeitsgruppe "Wolkenphysik und Radarmeteorologie" danke ich für die schöne Zeit auch außerhalb des Büros, z.B. beim Billardspielen, Kegeln, Grillen und Glühweintrinken. Ich würde mich freuen, wenn wir solche gemeinsamen Aktivitäten auch nach dem Ende der Arbeitsgruppe ab und zu wiederholen könnten (falls es mich noch eine Weile in Karlsruhe hält).

Stephanie Fiedler war mir, wenn auch nur für zwei Monate, eine freundliche Bürokollegin. Die gemeinsamen Kaffeepausen halfen mir den Kopf wieder frei zu kriegen.

Allen Mitarbeitern am IMK danke ich für die gute kollegiale Atmosphäre hier am Institut.

Ich möchte nicht vergessen mich auch bei Daniel Leuenberger zu bedanken, der mit seiner Vorarbeit zu Beginn des Projekts einen wesentlichen Beitrag leistete und uns den Einstieg in die Entwicklung des Radarvorwärtsoperators erleichterte.

Monika Pfeifer und Martin Hagen gaben mir interessante Einblicke in die Simulation von Polarisationsparametern, die leider aufgrund des zeitlichen Rahmens in meiner Arbeit nicht mehr umgesetzt werden konnte, von dem Wissen konnte meine Arbeit dennoch profitieren.

Der Deutschen Wetterdienst machte mit seinen finanziellen Mitteln, die Umsetzung des Projekts erst möglich.

Mein früherer Kollege Felix Werner, der das Glück hatte eine Woche vor mir seine Promotion zu beenden, half mir mit wertvollen Tipps den organisatorischen Überblick und die nötige Ruhe zu bewahren.

Meinem Mann Markus Jerger danke ich für das gründliche Korrekturlesen meiner Arbeit und die Hilfestellung beim Erlernen von Python – auch von Australien aus.

Zu guter Letzt bedanke ich mich bei meiner gesamten Familie für die viele Unterstützung und das Interesse an meiner Arbeit.

»Und so sehen wir betroffen - Den Vorhang zu und alle Fragen offen.«

(M. Reich-Ranicki in Anlehnung an B. Brecht)

Licht aus.

Wissenschaftliche Berichte des Instituts für Meteorologie und Klimaforschung des Karlsruher Instituts für Technologie (0179-5619)

Bisher erschienen:

Nr. 1: *Fiedler, F. / Prenosil, T.*
Das MESOKLIP-Experiment. (Mesoskaliges Klimaprogramm im Oberrheintal).
August 1980

Nr. 2: *Tangermann-Dlugi, G.*
Numerische Simulationen atmosphärischer Grenzschichtströmungen über langgestreckten mesoskaligen Hügelketten bei neutraler thermischer Schichtung.
August 1982

Nr. 3: *Witte, N.*
Ein numerisches Modell des Wärmehaushalts fließender Gewässer unter Berücksichtigung thermischer Eingriffe.
Dezember 1982

Nr. 4: *Fiedler, F. / Höschele, K. (Hrsg.)*
Prof. Dr. Max Diem zum 70. Geburtstag.
Februar 1983 (vergriffen)

Nr. 5: *Adrian, G.*
Ein Initialisierungsverfahren für numerische mesoskalige Strömungsmodelle.
Juli 1985

Nr. 6: *Dorwarth, G.*
Numerische Berechnung des Druckwiderstandes typischer Geländeformen.
Januar 1986

Nr. 7: *Vogel, B.; Adrian, G. / Fiedler, F.*
MESOKLIP-Analysen der meteorologischen Beobachtungen von mesoskaligen Phänomenen im Oberrheingraben.
November 1987

Nr. 8: *Hugelmann, C.-P.*
Differenzenverfahren zur Behandlung der Advektion.
Februar 1988

Nr. 9: *Hafner, T.*
Experimentelle Untersuchung zum Druckwiderstand der Alpen.
April 1988

Nr. 10: *Corsmeier, U.*
Analyse turbulenter Bewegungsvorgänge in der maritimen atmosphärischen Grenzschicht.
Mai 1988

Nr. 11: *Walk, O. / Wieringa, J.(eds)*
Tsumeb Studies of the Tropical Boundary-Layer Climate.
Juli 1988

Nr. 12: *Degrazia, G. A.*
Anwendung von Ähnlichkeitsverfahren auf die turbulente Diffusion in der konvektiven und stabilen Grenzschicht.
Januar 1989

Nr. 13: *Schädler, G.*
Numerische Simulationen zur Wechselwirkung zwischen Landoberflächen und atmophärischer Grenzschicht.
November 1990

Nr. 14: *Heldt, K.*
Untersuchungen zur Überströmung eines mikroskaligen Hindernisses in der Atmosphäre.
Juli 1991

Nr. 15: *Vogel, H.*
Verteilungen reaktiver Luftbeimengungen im Lee einer Stadt – Numerische Untersuchungen der relevanten Prozesse.
Juli 1991

Nr. 16: *Höschele, K.(ed.)*
Planning Applications of Urban and Building Climatology – Proceedings of the IFHP / CIB-Symposium Berlin, October 14-15, 1991.
März 1992

Nr. 17: *Frank, H. P.*
Grenzschichtstruktur in Fronten.
März 1992

Nr. 18: *Müller, A.*
Parallelisierung numerischer Verfahren zur Beschreibung von Ausbreitungs- und chemischen Umwandlungsprozessen in der atmosphärischen Grenzschicht.
Februar 1996

Nr. 19: *Lenz, C.-J.*
Energieumsetzungen an der Erdoberfläche in gegliedertem Gelände.
Juni 1996

Nr. 20: *Schwartz, A.*
Numerische Simulationen zur Massenbilanz chemisch reaktiver Substanzen im mesoskaligen Bereich.
November 1996

Nr. 21: *Beheng, K. D.*
Professor Dr. Franz Fiedler zum 60. Geburtstag.
Januar 1998

Nr. 22: *Niemann, V.*
Numerische Simulation turbulenter Scherströmungen mit einem Kaskadenmodell.
April 1998

Nr. 23: *Koßmann, M.*
Einfluß orographisch induzierter Transportprozesse auf die Struktur der atmosphärischen Grenzschicht und die Verteilung von Spurengasen.
April 1998

Nr. 24: *Baldauf, M.*
Die effektive Rauhigkeit über komplexem Gelände – Ein Störungstheoretischer Ansatz.
Juni 1998

Nr. 25: *Noppel, H.*
Untersuchung des vertikalen Wärmetransports durch die Hangwindzirkulation auf regionaler Skala.
Dezember 1999

Nr. 26: *Kuntze, K.*
Vertikaler Austausch und chemische Umwandlung von Spurenstoffen über topographisch gegliedertem Gelände.
Oktober 2001

Nr. 27: *Wilms-Grabe, W.*
Vierdimensionale Datenassimilation als Methode zur Kopplung zweier verschiedenskaliger meteorologischer Modellsysteme.
Oktober 2001

Nr. 28: *Grabe, F.*
Simulation der Wechselwirkung zwischen Atmosphäre, Vegetation und Erdoberfläche bei Verwendung unterschiedlicher Parametrisierungsansätze.
Januar 2002

Nr. 29: *Riemer, N.*
Numerische Simulationen zur Wirkung des Aerosols auf die troposphärische Chemie und die Sichtweite.
Mai 2002

Nr. 30: *Braun, F. J.*
Mesoskalige Modellierung der Bodenhydrologie.
Dezember 2002

Nr. 31: *Kunz, M.*
Simulation von Starkniederschlägen mit langer Andauer über Mittelgebirgen.
März 2003

Nr. 32: *Bäumer, D.*
Transport und chemische Umwandlung von Luftschadstoffen im Nahbereich von Autobahnen – numerische Simulationen.
Juni 2003

Nr. 33: *Barthlott, C.*
Kohärente Wirbelstrukturen in der atmosphärischen Grenzschicht.
Juni 2003

Nr. 34: *Wieser, A.*
Messung turbulenter Spurengasflüsse vom Flugzeug aus.
Januar 2005

Nr. 35: *Blahak, U.*
Analyse des Extinktionseffektes bei Niederschlagsmessungen mit einem C-Band Radar anhand von Simulation und Messung.
Februar 2005

Nr. 36: *Bertram, I.*
Bestimmung der Wasser- und Eismasse hochreichender konvektiver Wolken anhand von Radardaten, Modellergebnissen und konzeptioneller Betrachtungen.
Mai 2005

Nr. 37: *Schmoeckel, J.*
Orographischer Einfluss auf die Strömung abgeleitet aus Sturmschäden im Schwarzwald während des Orkans „Lothar".
Mai 2006

Nr. 38: *Schmitt, C.*
Interannual Variability in Antarctic Sea Ice Motion: Interannuelle Variabilität antarktischer Meereis-Drift.
Mai 2006

Nr. 39: *Hasel, M.*
Strukturmerkmale und Modelldarstellung der Konvektion über Mittelgebirgen.
Juli 2006

Ab Band 40 erscheinen die Wissenschaftlichen Berichte des Instituts für Meteorologie und Klimaforschung bei KIT Scientific Publishing (ISSN 0179-5619).
Die Bände sind unter www.ksp.kit.edu als PDF frei verfügbar oder als Druckausgabe bestellbar.

Nr. 40: *Lux, R.*
Modellsimulationen zur Strömungsverstärkung von orographischen Grundstrukturen bei Sturmsituationen. (2007)
ISBN 978-3-86644-140-8

Nr. 41: *Straub, W.*
Der Einfluss von Gebirgswellen auf die Initiierung und Entwicklung konvektiver Wolken. (2008)
ISBN 978-3-86644-226-9

Nr. 42: *Meißner, C.*
High-resolution sensitivity studies with the regional climate model COSMO-CLM. (2008)
ISBN 978-3-86644-228-3

Nr. 43: *Höpfner, M.*
Charakterisierung polarer stratosphärischer Wolken mittels hochauflösender Infrarotspektroskopie. (2008)
ISBN 978-3-86644-294-8

Nr. 44: *Rings, J.*
Monitoring the water content evolution of dikes. (2009)
ISBN 978-3-86644-321-1

Nr. 45: *Riemer, M.*
Außertropische Umwandlung tropischer Wirbelstürme: Einfluss auf das Strömungsmuster in den mittleren Breiten. (2012)
ISBN 978-3-86644-766-0

Nr. 46: *Anwender, D.*
Extratropical Transition in the Ensemble Prediction System of the ECMWF: Case Studies and Experiments. (2012)
ISBN 978-3-86644-767-7

Nr. 47: *Rinke, R.*
Parametrisierung des Auswaschens von Aerosolpartikeln durch Niederschlag. (2012)
ISBN 978-3-86644-768-4

Nr. 48: *Stanelle, T.*
Wechselwirkungen von Mineralstaubpartikeln mit thermodynamischen und dynamischen Prozessen in der Atmosphäre über Westafrika. (2012)
ISBN 978-3-86644-769-1

Nr. 49: *Peters, T.*
Ableitung einer Beziehung zwischen der Radarreflektivität, der Niederschlagsrate und weiteren aus Radardaten abgeleiteten Parametern unter Verwendung von Methoden der multivariaten Statistik. (2012)
ISBN 978-3-86644-323-5

Nr. 50: *Khodayar Pardo, S.*
High-resolution analysis of the initiation of deep convection forced by boundary-layer processes. (2012)
ISBN 978-3-86644-770-7

Nr. 51: *Träumner, K.*
Einmischprozesse am Oberrand der konvektiven atmosphärischen Grenzschicht. (2012)
ISBN 978-3-86644-771-4

Nr. 52: *Schwendike, J.*
Convection in an African Easterly Wave over West Africa and the Eastern Atlantic: A Model Case Study of Hurricane Helene (2006) and its Interaction with the Saharan Air Layer. (2012)
ISBN 978-3-86644-772-1

Nr. 53: *Lundgren, K.*
Direct Radiative Effects of Sea Salt on the Regional Scale. (2012)
ISBN 978-3-86644-773-8

Nr. 54: *Sasse, R.*
Analyse des regionalen atmosphärischen Wasserhaushalts unter Verwendung von COSMO-Simulationen und GPS-Beobachtungen. (2012)
ISBN 978-3-86644-774-5

Nr. 55: *Grenzhäuser, J.*
Entwicklung neuartiger Mess- und Auswertungsstrategien für ein scannendes Wolkenradar und deren Anwendungsbereiche. (2012)
ISBN 978-3-86644-775-2

Nr. 56: *Grams, C.*
Quantification of the downstream impact of extratropical transition for Typhoon Jangmi and other case studies. (2013)
ISBN 978-3-86644-776-9

Nr. 57: *Keller, J.*
Diagnosing the Downstream Impact of Extratropical Transition Using Multimodel Operational Ensemble Prediction Systems. (2013)
ISBN 978-3-86644-984-8

Nr. 58: *Mohr, S.*
Änderung des Gewitter- und Hagelpotentials im Klimawandel. (2013)
ISBN 978-3-86644-994-7

Nr. 59: *Puskeiler, M.*
Radarbasierte Analyse der Hagelgefährdung in Deutschland. (2013)
ISBN 978-3-7315-0028-5

Nr. 60: *Zeng, Y.*
Efficient Radar Forward Operator for Operational Data Assimilation within the COSMO-model. (2014)
ISBN 978-3-7315-0128-2

Nr. 61: *Bangert, M. J.*
Interaction of Aerosol, Clouds, and Radiation on the Regional Scale. (2014)
ISBN 978-3-7315-0123-7

Nr. 62: *Jerger, D.*
Radar Forward Operator for Verification of Cloud Resolving Simulations within the COSMO Model. (2014)
ISBN 978-3-7315-0172-5